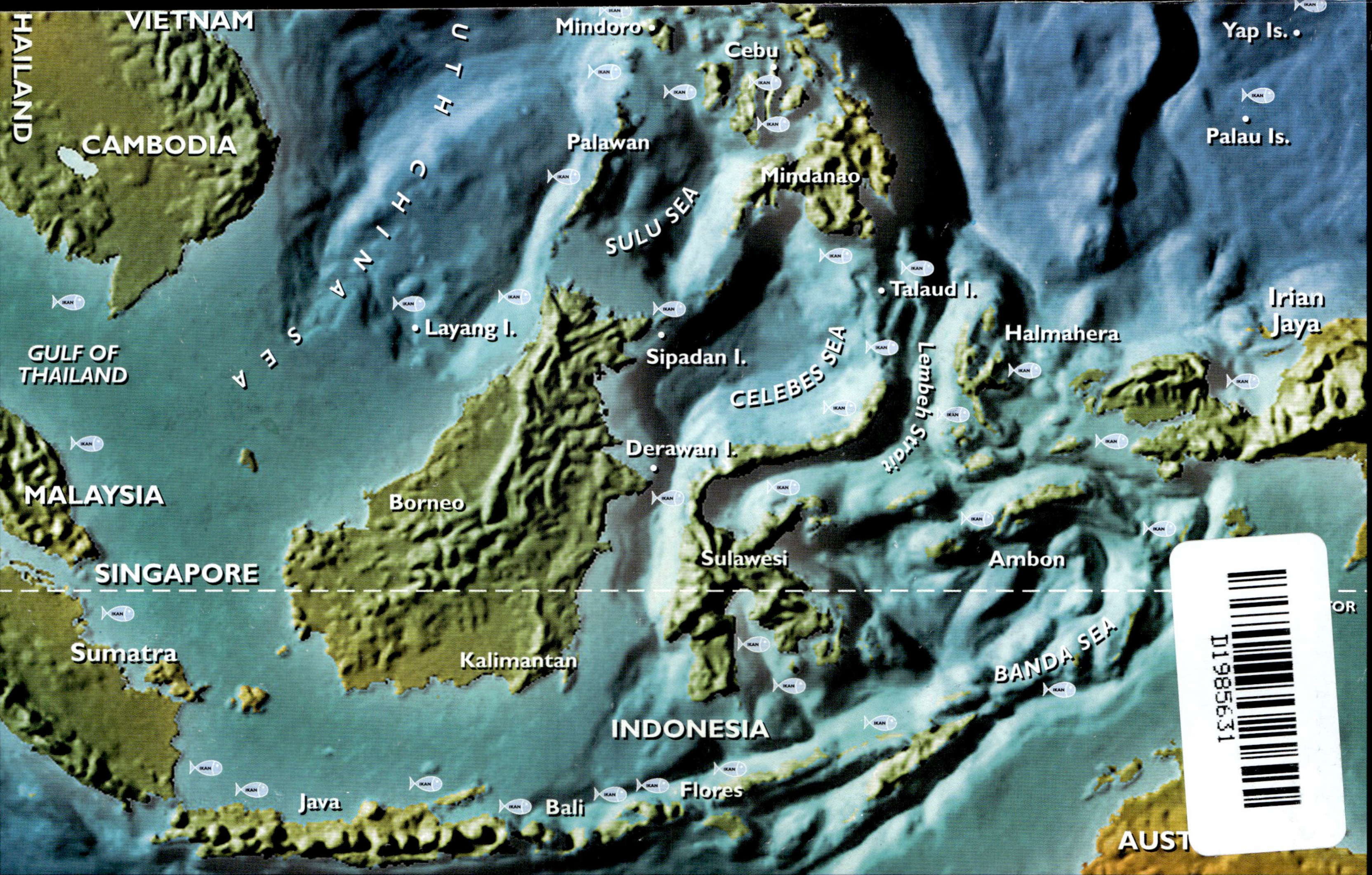
VIETNAM
HAILAND
CAMBODIA
GULF OF THAILAND
MALAYSIA
SINGAPORE
Sumatra
Java
Bali
Flores
UTH CHINA SEA
Layang I.
Mindoro
Palawan
Cebu
Mindanao
SULU SEA
Sipadan I.
CELEBES SEA
Derawan I.
Borneo
Kalimantan
Sulawesi
INDONESIA
Talaud I.
Lembeh Strait
Halmahera
Ambon
BANDA SEA
Yap Is.
Palau Is.
Irian Jaya
AUST
OR

First edition published in Germany 2001
and to be ordered directly by
IKAN - Unterwasserarchiv
Waldschulstrasse 166
65933 Frankfurt, Germany
Fax: (Int. 49) 69-383587 - e-mail: ikanuw@aol.com

Conception & Layout: Helmut Debelius
Type-setting & Translation: Ralf Michael Hennemann
Print & Production: Grupo M&G Difusión, S.L.

Helmut Debelius

ASIA PACIFIC REEF GUIDE

Malaysia • Indonesia • Palau
Philippines • Tropical Japan • China
Vietnam • Thailand

Over 1,000 photographs
of coral reef animals
taken in their natural habitat

IKAN

TABLE OF CONTENTS 2 - 5

PICTURE STORIES

PHOTO CREDITS

a = all, t = top, c = centre, b = bottom, w = whole p.
GERRY ALLEN 136c
CHARLES ANDERSON 270c 277t b
JIM BLACK 279b 281t
CLAY BRYCE 273b
HELMUT DEBELIUS 11t t b 12t c 13c 15t b 17a 18b 23b 32b 33t 34a 35t c 36b 40t c 42t 45c 46t 47a 49b 50b 52t 53t c 63a 64c 68a 71b 76c 77b 78c b 79a 80t 81a 83t b 84a 85c b 86c 87t c 92b 93a 95t c 96a 98c 99c b 100c b 101c b 106c b 107c b 108t c 111t b 112c b 113a 114t 115b 116a 120c 121t c 122c b 123t 124t b 125t 126c b 127t c 128t c 129t b 134c 137a 138t 139t b 140c b 143b 144c 145t 146a 147b 148c 149t b 152b 153t 155t b 156b 158t 159w 161a 162a 163t 164c 166b 167t b 170b 171t b 172a 176c b 177t b 179c 180a 181t 182c 186a 187b 188t 189b 190t 191t c 192c b 193b 194b 201b 202a 203c 204b 205c b 206t 207t c 208c b 209b 210b 216c b 217b 218t b 219c t 220b 221c 222b 223a 224t c 225t b 226t b 227t b 228t b 229c b 232a 233b 234a 235t c 236t 237t c 238b 239a 242t 243a 248b 249t c 250c b 251c b 252t 253t b 254t 255t b 258c b 259b 260a 261t 262a 266b 267t b 268b 269t c 270t b 271t 272a 273c 274t 275b 276c b 278t b 279c 280a 281c b 282c 283c b 284t b 285c 290c b 291a 292a 293t b 294t b 295w 296b 297b 298tl tr b 299tr b 300a 301a 302a 303a 304a 308b 309c b 310a 311a 312a 313a 314a 315t b
BOB HALSTEAD 22b 24t 268c; TOMONORI HIRATA 61tr; HARALD JANUSCHKE 261b; AVI KLAPFER 271c; SABINE KLEIN 241c; DANIEL KNOP 67b 267c 268t; PETRA KÖLLE 97w
RUDIE KUITER 13t b 16c b 18c 19t 24c b 25t b 32t 33c b 35b 36c 37a 38a 40b 41t 42c b 45b 46c 48c b 50t 52c 60t b 62a 64t b 76t b 78t 80c b 82c 85t 86t b 87b 94a 98b 100t 101t 110c 111c 114c b 120b 121b 123c b 124c 125c b 127b 128b 129c 136t 142c b 143c 144t b 145c 147c 148b 152c 153b 154t b 155c 156t c 157c b 163b 164t b 165a 167c 170c 173t b 177c 178a 179t b 181c b 182b 187t 188c b 190c b 192t 194t 201t c 203t b 204t c 206c b 207b 209t 210t c 219b 220t 221t b 225c 226c 235b 237b 238t 240c b 241t b 271b 299tl
KEVIN MARKEY 19b; MARK PREIN 279t
DOUG PERRINE 135w 138b 143t 200b 242b
ED ROBINSON 10b 20b 22t 49c 50c 109w 141w 160b 166t 249b 269b 285t b 294c 296t 297t 309t
SHIMLOCK/JONES 19c 23c 51w 52b 58b 60c 61b 69w 70t c 71t c 99t 189c 251t 254b 273t 276t 308t
ROGER STEENE 49t 98t 208t 217t 220c 250t 253c 259c 263t
MARK STRICKLAND 14b 20t 21w 59w 193t c 217c 274c 275c 278c
LADAWAN SUPHANTHARIDA 258t
DENISE TACKETT 41b
HIROYUKI TATSUUMA 9w 12b 14t c 16t 22c 46b 48t 70b 82t b 95b 107t 110b 142t 145b 147t 158c b 182t 187c 209c 218c 222t 224b 229t 236c b
TAKAMASA TONOZUKA 39b 53b 61tl 205t 259t 275t 277c 282t b 283t 284c

PHOTO CREDITS COVER, left to right:
Symphorichthys spilurus - HELMUT DEBELIUS, *Reticulidia halgerda* - NORBERT PROBST, *Amphiprion ocellaris* - HELMUT DEBELIUS, *Synchiropus splendidus* - MARK STRICKLAND, *Stonogobius* sp. - TOMONORI HIRATA, *Spondylus varius* - HELMUT DEBELIUS

BACK: *Platax pinnatus* - TAKAMASA TONOZUKA

Introduction

This book replaces the Southeast Asia Tropical Fish Guide (also published by IKAN), which is out of print since two years. However, it features some significant changes: Since the Andaman Sea in the eastern Indian Ocean is covered by the new INDIAN OCEAN REEF GUIDE, the present volume concentrates on Pacific waters only. The geographical area of the ASIA PACIFIC REEF GUIDE therefore includes Malaysia and Indonesia in the south, the Gulf of Thailand and the Vietnamese and Chinese coasts in the west, and in the north all islands of tropical/subtropical Japanese waters in the East China Sea south of 30°N latitude. It also covers the habitats of popular diving destinations like the oceanic islands of Palau, Yap, or Ogasawara in the east. Furthermore it includes not only fishes but also a selection of invertebrates. Rather than including a long textbook-like list of lower animals, I have consciously focused on those groups that are more likely to attract the interest of divers and snorkellers. Over the years, I have had sufficient opportunities to become familiar with their particular wishes.

SHIMLOCK/JONES

To understand more about the animals' behaviour and also ecological aspects, the ASIA PACIFIC REEF GUIDE includes 20 additional picture stories from this region and written by various underwater photographers and myself. Some of these stories clearly show that the proverbial 'healthy' underwater world does not exist anymore in many parts of this region because many fishes and invertebrates are subject to an enormous 'predation pressure' by man. By documenting the techniques of removing edible marine organisms

Above: Fantastic diving locations entice divers from all over the world to come to the Asian Pacific region. Right: Not only divers feel at home there but also Orang Utans...

from coral reefs I do not intend to offend anybody, though. I simply want to alert every reader of this book and to point towards the long-term effects of such destructive practices. Dive tourism which is based on healthy reefs with a full spectrum of marine life will be far more rewarding on the long run to the local communities in the countries concerned.

According to the author diving is best off a live-aboard like the one above.

There was no need to include elasmobranchs in this book since the recently published IKAN-volume SHARKS & RAYS GUIDE OF THE WORLD - avoiding the usual drawings but illustrating 240 species with underwater photos - became the most comprehensive popular marine guide on this group at present. To the fans of these major predators of the sea I only can recommend to have a look into this unique book with breathtaking illustrations including also reef-dwelling sharks and rays. However, the pending question remains: Will divers encounter sharks in the Asia Pacific region in the future? Fishing, in particular the practice of shark finning, has seriously and wastefully depleted shark and ray populations. Many species are targetted directly, while others are discarded dead or dying as bycatch of numerous fisheries. So seriously damaging have these activities become, that some elasmobranch species have been almost eliminated in parts of their range.

The Asia Pacific region includes over 20,000 islands, countless coral reefs, and in addition there is a mixture of faunas from different oceans. These are some of the ingredients supporting the (still) richest marine fauna of the world. Just reef fishes alone are estimated to comprise more than 3,000 species living in our area. Underwater faunas can differ greatly between different areas, including from one island or indeed from one reef to the next. Just like on land where water or

In Asian Pacific waters one can watch wonderful fishes by day and at night: Swimming angel fish (left), sleeping rabbit fishes (right).

mountains divide the fauna, in the underwater realm currents, depths, and land-masses separate faunas just as effectively. Over time the water levels changed significantly, especially during the ice ages, dropping more than 100 m during certain periods which means that you could walk from China to Bali, and in addition temperatures also altered. This effected the fauna in several ways. During the times when the sea level was very low, a once united fauna became divided between the Indian and Pacific Oceans what caused changes in many of the species. When temperatures change, the fauna migrates to the adjacent areas that become more suitable, and that can be in several directions. On large land masses migration is usually possible as temperatures change, the fauna can remain in touch with the right conditions, but a small isolated island offers no alternatives other than adapting. This is one of the main reasons for the low species diversity on remote outcrops. The same factor has also caused colour variations in wide-ranging species, especially between populations along the continental margins and those more oceanic.

Further north in the area covered by this volume, the marine faunas are continuations along the continental margins, island chains, and ridges which embrace a number of seas. The South China Sea ranges from north of the Philippines along Vietnam and extends south to Malaysia and Singapore. The fauna along its continental margin ranges from the East China Sea (Southern Japan) to Java. The Philippine fauna is more oceanic with similar species ranging along the east from southern Japan to New Guinea. Between the South China Sea and the Pacific Ocean there are some large water masses: The Sulu Sea, Celebes Sea, and one of the most diverse areas, the Banda Sea, which is reaching down to a depth of 7,440 m.

Now after 25 years of active diving in the Asia Pacific region and witnessing a still rapidly growing sport, the knowledge about fishes has come to a point where we have a fairly good idea of what is there. However, there are many new species, and also similar species which create some confusion concerning their correct identity, and no doubt more will be found in the future. Some species were only recently described and a large number are in the process of being described. Identification work never finishes, and - more importantly - we definitely need to know more about the behaviour, ecology, habitats etc. of the species, and thus about the marine environment as a whole.

Frankfurt,
autumn of 2001

Helmut Debelius

STEVE DROGIN

Nowadays, it is not always guaranteed to dive in a healthy reef environment due the damage to the reef organisms caused by global warming. However, the author was lucky to dive many beautiful reefs in Asian Pacific locations like this one in Indonesia.

A large family of marine eels, comprising at least 10 genera and well over 100 species (200 by some authors), distributed world-wide in all tropical and subtropical seas. They feature muscular robust posteriorly compressed bodies, with tough thick skin and a single median fin present, usually originating dorsally before gill opening, continuing around tail to anus. Mouth up to moderately large size, with variable teeth between genera. Gill openings reduced, as a small hole, mid-laterally behind head. Most species have distinct coloration, consisting of spots and stripes, often with complicated patterns which are usually typical for a species. Mostly reef dwellers, living secretive in holes and crevices. Usually observed with only the head exposed, however a few species move about on reefs in the open hunting for food. They feed on a variety of invertebrates and some species feed on fish. Although generally not aggressive, the larger species can inflict serious and painful wounds. Some have relatively poor eye-sight and rely more on smell, often coming towards a diver to investigate. Sometimes they congregate in certain areas apparently to spawn, and in such areas they are known to attack divers.

Our knowledge of breeding behaviour is limited to a few species seen intertwine and release pelagic eggs. Some eels are known to migrate over great distances but most tropical reef eels remain in their home territory or congregate in small groups in some areas. Larvae have deep greatly compressed and clear bodies, settling at about 50 mm length as they change shape to the more rounded as in the adults. As adult they range in size from about 30 cm to 2 m.

Rhinomuraena quaesita **Palau**

Ribbon eel

Length: To 120 m.
Distribution: Indo-Pacific, including all of our area.
Depth: 1-55 m.
General: The most distinct moray of tropical coral reefs. The secretive species lives hidden in sand or among coral rubble with only the head showing. Head and part of the very long body reach out of the burrow to grab its prey of small fishes (e.g. damselfishes). Coastal reef crests and slopes, usually seen singly but occasionally a pair or two females may share the same burrow. Left and below: Unusual photos since Ribbon eels are rarely seen outside their holes!
Continued on next page.

Ribbon eel
Continued.

The species is a protandrous hermaphrodite which means the individuals reach sexual maturity as males and later change their sex to become females. During this transition they also change their coloration. Attention: Latest scientific research states that males are black (immediately right), females are brilliantly blue with bright yellow fins (extreme right), and old females become uniformly yellow (small photo below, from Pulau Seribu, Indonesia).

R. quaesita **Flores, Indonesia** **Cebu, Philippines**

Honeycomb moray

Le: To 200 cm. Di: Throughout tropical Indo-Pacific, including our area. De: 3-50 m. Ge: One of the largest morays, easily identified by the coloration. Coastal protected shallow bays to deep reefs. Rare in the eastern parts of west Pacific. *G. isingteena* (below) is similar.

Gymnothorax favagineus **Kalimantan**

Gymnothorax javanicus **Luzon, Philippines**

Giant moray

Length: To 240 cm.
Distribution: Indo-Pacific, throughout our area.
Depth: 10-50 m.
General: One of the largest morays which has the potential to inflict injury to divers if they are not careful enough when feeding other fish, as has been practised in some tourist operations. The species is found in various reef habitats from coastal to outer reefs.

Gymnothorax flavimarginatus **Banda Sea, Indonesia**

Yellow-edged moray

Length: To 120 cm.
Distribution: Indo-Pacific, including all of our area.
Depth: 3-100+ m.
General: A common species on most reefs, especially on mixed rock and coral reefs. The species participates readily in shark or grouper feeding events and is able to inflict nasty wounds to inexperienced divers.

Gymnothorax breedeni **Okinawa, Japan**

Masked moray

Length: To 100 cm.
Distribution: Uncommon but found throughout our area.
Depth: 4-25 m. General: This moray prefers oceanic and seaward reefs, sometimes sharing its lair with a partner. Easily identified by black mark from eye to rear of mouth.

Undulate moray

Length: To 100 cm (reported to 150 cm).
Distribution: Indo-Pacific, including all of our area, wide ranging from east Africa to Panama.
Depth: 5-50 m.
General: A common species, mostly observed at night when it hunts by moving around in search for small fish or shrimps. In Bali this species was photographed taking a cardinalfish from a hole in a sponge. Found on coastal reef slopes, this species can be aggressive towards divers.

Gymnothorax undulatus **Flores, Indonesia**

Spot-face moray

Le: To 40 cm. Di: Indonesia to Philippines, southern Japan, and Micronesia. De: 10-50 m. Ge: Only known to be common at Bali, elsewhere reported as rare. At Bali often at the steep sand slopes with numerous isolated patch reefs and sponges in which they make their home.

Gymnothorax fimbriatus **Sulawesi**

Bar-tail moray

Le: To 46 cm. Di: Throughout area. De: 6-30 m. Ge: Seems to prefer reefs with rich coral growth where it hides during the day and moves about in the open at night. Distinctly coloured but only by the pattern of the posterior part of the body.

Gymnothorax zonipectis **Sulawesi**

Siderea picta Yonaguni I., Japan

Painted moray
Le: To 68 cm. Di: Indo-Pacific, including all of our area. De: 0-12 m. Ge: Shallow, may leave the water at low tide to hunt crabs. Below: **White-eyed moray** *S. thyrsoidea*, to 65 cm, throughout our area, 1-35 m. Coastal, silty habitats, reefs, lagoons. Eyes distinctly white. Common, often in pairs.

Gymnothorax eurostus Ogasawara Is., Japan

Spotted moray

Length: To 40 cm.
Distribution: West Pacific, including all of our area.
Depth: 2-25 m.
General: Found in crevices of coral reefs. Numerous small yellowish spots that gradually become larger on the posterior part of the body are distinctive for this species. Dark spots and blotches are primarily seen on the anterior half of this moray's body.

Gymnothorax melatremus Luzon, Philippines

Golden moray

Length: To 26 cm.
Distribution: Indo-Pacific, including all of our area.
Depth: 5-26+ m.
General: A small rare moray that is found in coastal bays on sponges and coral reefs and known from a few locations only. With its bright coloration it should be easily seen when occurring in habitats in our area. The teeth of this species are short and conical.

White-mouth moray

Length: To 100 cm.
Distribution: Indo-Pacific, throughout our area.
Depth: 1-35 m.
General: Found on clear coastal and protected inner reefs, usually living amongst corals in rich mixed coral and algae habitats. Best distinguished from similar moray species by the white inner lining of its mouth.

Below: **Yellowmouth moray** *G. nudivomer,* to 180 cm, Indo-Pacific, throughout our area, 4-165 m. Wide ranging species, inhabits crevices singly or in pairs. Also active during the day, leaving its lair. Inner lining of mouth brightly yellow.

Gymnothorax meleagris **Sangihe I., Indonesia**

Dragon moray

Length: To 80 cm.
Distribution: Indo-Pacific, including all of our area.
Depth: 15-50 m.
General: A relatively rare species and only commonly observed in southern Japan and the northern Philippines, but reported to northern Australia and various locations in the Indian Ocean. Appears to prefer rocky, rather than coral-dominated reefs.

Enchelycore pardalis **Cebu, Philippines**

Gymnothorax enigmaticus **Okinawa, Japan**

Tiger moray

Length: To 58 m.
Distribution: Indo-Pacific, including all of our area.
Depth: 16-23 m.
General: This moray is relatively rare and found mainly on reef flats in the intertidal zone, in some regions also on offshore reefs. An important feature for identifying the species correctly are its black bands at the front of the body which are not interrupted on the underside (as present in the closely related and sympatrically living *G. ruepelliae)*.

Echidna nebulosa **Bali, Indonesia**

Clouded moray

Le: To 70 m. Di: Indo-Pacific, including all of our area. De: 1-15 m. Ge: Coastal and very shallow reef flats from the intertidal zone to lagoons on inner reefs. Sometimes found under rocks in pools at low tide. May hunt in bright sunlight in shallow lagoons, moving about in the open.

Echidna polyzona **Kerama I., Japan**

Barred moray

Length: To 60 cm.
Distribution: Indo-Pacific, including all of our area.
Depth: 1-15 m.
General: A shallow reef species found on clear coastal flats and lagoons. Secretive, hunting under the cover of the reef in search for small fishes and shrimps. Opportunistic feeder, but mostly observed only at night because shady places provide cover for this moray during the day.

White ribbon moray

Length: To 105 cm.
Distribution: Indo-Pacific, throughout our area.
Depth: 10-30 m.
General: This moray inhabits shallow reef flats, sheltered coastal reefs, and lagoons. It lives on sand or rubble bottoms with rocky outcrops and spends most of the time burrowing in the ground. It is easily recognised by its overall pale coloration and compressed ribbon-like body but not often encountered by divers even during the night hours when it comes out of the substrate to feed.

Pseudechidna brummeri **Gulf of Thailand**

Whitemargin moray

Le: To 120 cm. Di: All of our area. De: 0-25 m. Ge: A rare moray, can be found while reef walking at low tide. Below: **Tidepool snake moray** *Uropterygius micropterus,* to 28 cm, throughout our area, 0-25 m. On rubble on intertidal reef flats. Shares habitat with other small morays.

Enchelycore schismatorhynchus **Flores, Indonesia**

Zebra moray

Length: To 150 cm.
Distribution: Indo-Pacific, throughout our area.
Depth: 2-50 m
General: Coastal to outer reef crests and slopes. Very secretive, virtually only seen at night when they hunt the reefs for small fishes and shrimps. Easily identified with the distinct banding.

Gymnomuraena zebra **Sumatra**

A very large family of marine eels, comprising over 50 genera and 250 species world-wide. Highly diverse group, the species have moderately robust to extremely long tubular bodies. The head is usually sharply pointed and often the body ends in a distinctly pointed tail. The species have various colour patterns from uniformly pale to spotted or conspicuously banded. Usually these eels are buried in the substrate (sand or mud) and rarely seen during the day. A few banded species move about in the open, usually in shallow coastal areas, resembling venomous sea snakes. Most species feed on small fishes and crustaceans. Size of adults varies greatly between species, ranging from about 20 cm to 2.5 m.

There are a great number of species in the waters of our area, but because of their secretive habits even the most common species are only rarely seen. What can be usually seen of the burying species is merely the head which still is well camouflaged when peering above the substrate. Included here are the species occasionally seen out in the open while they hunt for small prey.

Myrichthys colubrinus **Bali, Indonesia**

Banded snake eel

Length: To 97 cm.
Distribution: Indo-Pacific, throughout our area.
Depth: 1-25 m.
General: This species is often seen by snorkellers and usually mistaken for a sea snake. It mimics the venomous sea snakes found living sympatrically in many areas and the colour pattern of the species varies accordingly. In Japan a banded sea snake eating the banded snake eel was photographed. The Pacific form is usually fully banded but Indian Ocean populations are often half-banded or have round spots alternating with bands.

The photos show different colour variants.

Myrichthys colubrinus **Gulf of Thailand**

Spotted snake eel

Le: To 100 cm. Di: Indo-Pacific, throughout our area. De: 1-50 m. Ge: Size and number of spots highly variable, changing from few and large to many and small in adults. Common, shallow coastal sand flats, reef crest, drop-offs and lagoons, mainly out hunting at night. Below: A successful hunter.

Myrichthys maculosus **Kerama I., Japan**

Crocodile snake eel

Length: To 125 cm.
Distribution: Indo-Pacific, throughout our area.
Depth: 1-30 m.
General: Usually seen with head peering from sand or loose rubble bottom, only rarely is the whole eel visible. Eyes further forward than in the very similar sympatric species *B. cirrocheilos*. Ambushes small fishes and crustaceans. Colour is variable. The photo shows a specimen that has just swallowed a Flutemouth or Cornetfish *(Fistularia* sp.).

Brachysomophis crocodilinus **Lembeh Strait, Indonesia**

Dark-shouldered snake eel

Length: To 90 cm.
Distribution: West Pacific, throughout our area.
Depth: 7-35 m.
General: Found in coastal mud and sand habitats. Easily recognised by distinctive colour pattern of white-edged black blotch centrally on head. Preys on small fishes that are caught in the open at night.

The unusual photo shows the snake eel together with the moray *Echidna nebulosa* (see p.16).

Ophichthus cephalozona **Mindoro, Philippines**

Ophichthus bonaparti **Lembeh Strait, Indonesia**

Clown snake eel

Length: To 70 cm.
Distribution: Indo-Pacific, throughout our area.
Depth: 1-30 m.
General: This attractive snake eel lives in sand and sea grass beds and is mostly seen with just its head peering from the sand. Occasionally the whole eel is found wandering along the bottom. When disturbed it uses its tail to dig into the sand and can rapidly disappear. It does not build a permanent burrow. The banded body mimics that of sea snakes and probably provides some protection, however the tail is pointed unlike the paddle tail of sea snakes and therefore easy for divers to distinguish.

Callechelys marmorata **Palau**

Marbled snake eel

Length: To 65 cm.
Distribution: Indo-Pacific, throughout our area.
Depth: 1-30 m.
General: Easily recognised widespread but uncommon snake eel found in sand patches near reefs. As with most snake eels it is more active at night. Spends time completely buried in sand. Much of our knowledge about snake eel identification comes from specimens collected incidentally when using poisons to collect fishes.

Facing page:
Black-finned snake eel
Ophichtus melanochir, to 80 cm, Indo-Pacific, including all of our area, 0-25 m. Found on sand bottoms in the vicinity of coral reefs. Easily recognisable by colour pattern of dorsal fin having a distinctively black edge, also pectoral fins of some specimens are entirely or partly black. Photo from Lembeh Strait, Indonesia.

Conger cinereus **Palau**

Moustache conger

Length: To 80 cm.
Distribution: Indo-Pacific, throughout our area.
Depth: 10-90 m.
General: Seems to be most common on reef flats and in seagrass beds of shallow lagoons but also occurs on the outer reef slopes. A nocturnally active predator hunting fishes and small crustaceans. Divers encounter this conger eel only rarely during the day when it is resting in rock crevices, sometimes in aggregations of several individuals. During the night it shows a pattern of broad dark bars.

Conger japonicus **Ogasawara Is., Japan**

Japanese conger

Length: To 80 cm.
Distribution: Temperate and subtropical West Pacific.
Depth: 15-80 m.
General: This is a species which is rarely photographed in tropical waters and which is shown here for the first time in a popular tropical identification guide.

Conger eels are readily distinguished from the otherwise similarly shaped and sized morays by the presence of small but distinct pectoral fins. Morays don't have any pectorals.

Poeciloconger fasciatus **Papua New Guinea**

Barred sand conger

Length: To 60 cm.
Distribution: Indo-Pacific, probably throughout our area but records are extremely rare.
Depth: 2-25 m.
General: May be found free swimming close to the bottom or buried with only head emerging from sand. Prefers loose sand habitat. Only a few specimens have been discovered but widespread from Madagascar to Hawaii, a single record from Sulawesi.

The garden eels are sometimes also regarded as a subfamily of the large conger-eel group, the latter mostly found deep, in muddy habitats, or in temperate zones, and are rarely encountered on or near reefs. The garden eels, however, are an interesting group of fishes for divers, especially as the name implies they are found in such great numbers together that it looks like a garden of eels. The group is under revision and several species are in the process of being described.

The members of this family typically are seen on open sand patches or flats within or well away from reefs, with head upwards and part of the body vertically protruding from the bottom. Some species are extremely long and thin, reaching above the substrate to feed on zooplankton drifting by. Usually these eels occur in dense colonies and large numbers, sometimes in a seemingly endless field. Most species live in areas with strong currents influenced by the tides and feed when the currents bring the plankton food either from outflowing lagoons or incoming oceanic waters, depending on the type of plankton organisms living there.

Spaghetti garden eel

Length: To about 50 cm.
Distribution: Indo-West Pacific, throughout our area.
Depth: 25-40 m.
General: A common species on sand flats in the vicinity of coral reefs. Often encountered as large colony with numerous members that rise high out of the substrate to feed on plankton that drift by in the current.

Photo below from Mabul Island, Malaysia.

Gorgasia japonica **Sulawesi**

Gorgasia japonica **Gulf of Thailand**

Heteroconger preclara — Papua New Guinea

Splendid garden eel
Le: To 60 cm. Di: Indo-Pacific, all of our area. De: 5-40 m. Ge: Splendidly coloured, easily recognised even from a distance. On sand patches near reefs. Difficult to approach.

Heteroconger perissodon — **Flores, Indonesia**

Many-toothed garden eel

Length: To 60 cm.
Distribution: Philippines to Coral Sea.
Depth: 2-50 m.
General: Lives in a wide range of habitats from shallow sea grass to deep sand slopes, easily distinguished by white blotch on side of head. As with all garden eels lives in colonies, favours sites with current and feeds on plankton drifting past its hole in the substrate. Disappears into hole when approached. The white blotch laterally on the head is distinctive.

Heteroconger hassi — **Bali, Indonesia**

Spotted garden eel
Le: To 40 cm. Di: IP, all of our area. De: 6-45 m. Ge: On large open sand patches between reefs and on slopes. Usually in 15+ m but found in the shallows if currents provide food.

Zebra garden eel

Length: To 40 cm.
Distribution: Coral Sea to Japan.
Depth: 1-10 m.
General: On shallow coastal mud flats, usually in medium-sized colonies. Comes out of its hole to feed on plankton during current runs. Diameter of body to about 15 mm. Colour pattern distinctive.

This garden eel has been encountered only in very shallow water close to shore above beds of sea grass. Other garden eel species are known to favour sheltered shallow sand areas close to shore and can survive as long as there are slight tidal currents that run along the beach.

This species is especially attractive on black lava sand.

Heteroconger polyzona **Flores, Indonesia**

Taylor's garden eel

Length: To 40 cm. Distribution: Indonesia, PNG, possibly elsewhere in our area. Depth: 5-40 m. General: Recently described species named after underwater film makers Ron and Valerie Taylor who discovered this garden eel in Milne Bay, PNG. Now known from various Indonesian diving locations. Usually in small groups on silty bottom. Sightings in other areas in the range of this book could be true records.

Heteroconger taylori **Bali, Indonesia**

THE SENSUAL SEA

Several studies have found that the average adult human thinks about sex during anywhere from 80-95% of waking minutes (and possibly higher during sleep). For adolescent males it might be more meaningful to attempt a figure for the number of seconds per year during which this subject does not occupy the brain. Shocking figures perhaps for purveyors of fire and brimstone, but no surprise to those who believe in evolution by natural selection. Underwater photographer Doug Perrine teaches a lesson in animal sex.

DOUG PERRINE

Sergeant major damselfish *Abudefduf vaigiensis* spawning. The female (with black bars) lays reddish eggs while the male (with blue bars) fertilises them. Photo from Bali, Indonesia.

Several studies have found that the average adult human thinks about sex during anywhere from 80-95% of waking minutes (and possibly higher during sleep). For adolescent males it might be more meaningful to attempt a figure for the number of seconds per year during which this subject does not occupy the brain. Shocking figures perhaps for purveyors of fire and brimstone, but no surprise to those who believe in evolution by natural selection. In the game of evolution, "he who dies with the most toys" does not win. Instead, the prize goes to he or she who dies with the most copies of his/her genetic code replicated in successive generations. Nearly all organisms more advanced than bacteria have selected sex as their primary and most effective method of accomplishing this goal (although some hedge their bets by utilising other methods of reproduction as well, depending upon circumstances). Simply speaking, sex is the most important advance in life on earth since the "invention" of multi-cellular organisms. Other milestones in the history of our planet, including the inventions of sliced bread, pizza, and the aqua-lung, pale by comparison. Sex is the "ingenious" method by which genes are re-shuffled, producing novel combinations, creating the variability which is the raw material upon which natural selection acts. Without genetic variability there can be no evolution, and without sex the only source of variability is mutation. Naturally-occurring mutations tend to happen at a rate of only once in every million or so replications - way too slow to produce (by themselves) the explosive adaptive radiations that filled the continents with dinosaurs, and later mammals, and filled the seas with 25,000 species of fish, and innumerable species of sexually-reproducing invertebrates, not to mention libidinous marine mammals and reptiles.

Sex, by blending half the genes from each parent to create novel combinations, creates a "smorgasbord" of individual variation, from which nature can "choose" the fittest combinations. This not only allows a species to become stronger or more fit, but, most importantly, allows it to adapt to changing environments. Humans, of course, have largely exempted ourselves from the process of natural selection, largely as a consequence of medical advances which have allowed us to reduce infant mortality to a level that's nearly meaningless in evolutionary terms. Ironically, this intervention will probably result in the extinction of thousands of other species long before it has any discernible effect on the genetic fitness of our own race. With another intervention, birth control (sadly not applied in equal proportion to the first), we have separated sex from reproduction, and turned it into a recreational pursuit. For most organisms, however, sex remains intimately linked to the processes of reproduction and evolution. In these terms sex is more important than food or water. It is more important than survival itself - for survival without reproduction means an evolutionary dead end, while reproduction followed by death leaves offspring which can continue the genetic line.

It is no wonder then, that many animals will abandon their normal caution, and even expose themselves to extreme danger in pursuit of sex. It is also no surprise that predators, including fishermen, have learned to take advantage of this indiscretion. Three-spot damselfish will rarely leave the shelter of the reef, except in early morning when they may dance to nearly a meter above the reef at the peak of a spawning loop. It is at precisely this moment that they are most likely to vanish into the waiting jaws of a trumpet fish. Male sea turtles are so reluctant to abandon a female, once they have mounted, that they will endure severe injury by other males attempting to displace them. Island fishermen have learned to capture pairs of mating turtles by swimming over to the blissful couple and tying a line around the female's flipper. The male will rarely release her while she is pulled over to the fishing boat. In some cases, the predators target the eggs, rather than the parents. Whale sharks appear out of the blue during mass spawnings of mutton snappers and corals, both to vacuum up the eggs, and to feed on other plankton which eat the eggs. For some organisms, including salmon and many species of octopus and squid, sexual reproduction is inevitably followed by death, allowing the parent to put all of its resources into the survival of the offspring, saving nothing for itself. After mating, with certain death on the way, a female octopus has no further use for a male, and may have one for her last meal, to give her additional strength to see the eggs through to hatching.

Given the strong sex drive that is necessary to insure adequate reproduction, it is also not surprising that some species which reproduce infrequently have, like humans, adapted the sex act to other purposes than reproduction. In the ocean, nowhere is this more obvious than among the dolphins, which appear to use sex for a number of social purposes. These include reinforcing social bonds, establishing dominance (in which case the sex may occur between two individuals of the same gender, and/or different species), and simple recreation. Certainly dolphins do not restrict their sexual activities to those which are considered morally acceptable by humans. Typical dolphin activities would, if performed by humans, be classed as gang rape, child molestation, incest, sodomy, bestiality, homosexuality, and of course promiscuity. Sex often occurs so casually that scientists have quipped that to a dolphin "it's like a handshake."

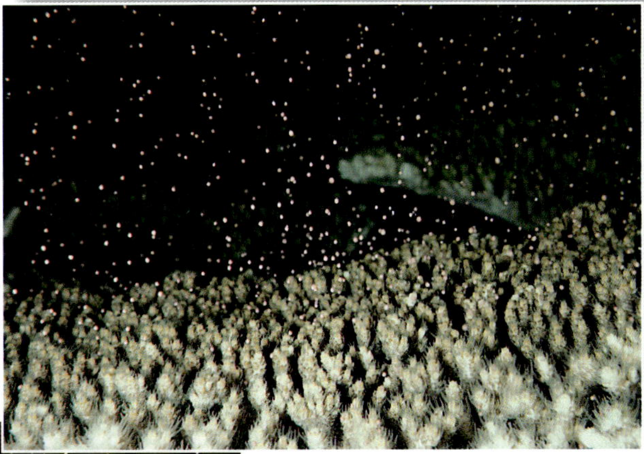

HELMUT DEBELIUS

Corals usually spawn at night. Shown here is an *Acropora* from Ko Samui, Gulf of Thailand. Despite the darkness butterfly fish are actively preying on the eggs.

This broad spectrum of sexual modes and partners seems to be common among marine mammals, including manatees, sea lions, and seals. Perhaps the most mysterious of these is the humpback whale. Both male and female humpbacks migrate to certain well-known breeding grounds where they give birth, and presumably mate. Even though several of these breeding areas are intensively studied, there is not a single reliable recorded observation of humpbacks mating. I have perhaps come as close as anybody to witnessing this arcane event. I have, on numerous occasions seen humpback whales engaged in what clearly appeared to be courtship. On three separate occasions, I have seen pairs of humpbacks, where one member of the couple had the penis extended 2-3m outside the body (where it is normally hidden when not in use). In two of these cases, however, both partners turned out to be male.

Given the vital importance of sex to evolution, and the many different organisms that employ it as their primary reproductive mode, we should not be surprised that evolution has produced many different variations on the sexual process itself, although these may indeed appear bizarre to humans who are not well-acquainted with the customs of the rest of the animal kingdom. In many marine organisms, gender is not as clearly defined as it is in our own kind. In turtles, crocodiles, some types of shrimp, and probably other kinds of animals as well, there are no "x and y" chromosomes which determine the sex of the offspring. Instead, the sex is determined by environmental factors during development. In sea turtles for example, the critical factor is the temperature of the nest. High temperatures produce all-female nests; low temperatures produce all males, and median temperatures produce mixed nests. By tampering with their reproductive efforts, for example by removing shoreline vegeta-

DOUG PERRINE

The colourful Pacific seastar *Pharia pyramidata* also spawns at night.

HELMUT DEBELIUS

tion or moving nests to shaded or unshaded hatcheries, we may be causing wild skews in the sex ratios of populations. For example, certain nesting beaches where eggs were collected and incubated in styrofoam coolers may have produced nothing but males for years.

Other creatures change gender during the course of their lives, or may be both male and female at the same time. Nudibranchs are simultaneous hermaphrodites, and engage both organs during sex, performing male duties at one end, and female duties at the other. Afterwards, both partners crawl off to become mothers, laying a delicate chiffon-like ribbon of eggs fertilised by the other nudibranch. Some flatworms, though also simultaneous hermaphrodites, apparently believe that it is more blessed to give than receive. Presumably their reproductive dynamics give males the edge in producing more offspring. When two individuals meet, they both rear up, evert their dagger-like penises and stab at each other. The winner of the duel punctures the loser's body wall and injects sperm, which fertilises the loser's eggs. It is believed that Freud had no knowledge of this ritual when he formulated his theories about "penis envy."

Some members of the grouper family are simultaneous hermaphrodites when young, but switch to male only later in life. Sex can get really complicated with some of these fish. In some cases the males maintain harems of hermaphrodites. The hermaphrodites mate only as females with the harem master, but may also mate as males with each other. In other cases the harem master is also a hermaphrodite, which mates only as a male with its concubines, but may take a "walk on the wild side" outside its territory to mate as a female with the harem master in another territory.

Courtship and reproduction of *Centropyge loricula* take place at dusk. Talaud Isl., Indonesia.

Many parrotfishes and wrasses are also protogynous sequential hermaphrodites. The terminal phase is often a "supermale" (so named because it is larger, more colourful, and dominant to initial phase fish, which may also include males). However sexual dynamics are much more complicated than a simple girl-to-boy progression, and vary from species to species. Supermales are often sex-reversed females, but in some species initial phase males may also progress to terminal phase. In some species there are several alternative life history patterns available. One of the better-studied is the tropical blue-headed wrasse. Only terminal-phase males have blue heads, while initial phase males and females and juveniles are all yellow. Supermales dominate spawning, as they are able to maintain spawning territories along the perimeter of the reef, which attract females. Initial phase males participate in group spawning and also sometimes spawn as "sneakers" or "streakers" - dashing in to release sperm just as a female is spawning with a supermale. Initial phase males, however, have lower reproductive success than terminal males. The number of available territories is a function of the length of the circumference of the reef, and the number of terminal males is in direct proportion to that figure. The overall ratio of males to females in each population is also a function of the size of the reef, and can be mathematically show to optimise the reproductive potential of the population. Obviously the sexual identity of these fish is being environmentally determined, but the mechanism is unknown. In other wrasse species, gender is socially determined - it depends upon the sex ratio already present in the population.

RUDIE KUITER

The true beauty of dragonfish is seen only during courtship. The male of *Synchiropus moyeri* displays his dorsal fin to attract the female in the background. Soon both together head for open water to release spawn and eggs. Photos from Bali, Indonesia.

In some organisms, the two sexes start out as different individuals, but merge into one body. In such disparate groups as deep-sea anglerfish, spoon worms, and parasitic barnacles, the male starts out life as a free-swimming larva, but follows a chemical trail to the female, enters her body, and is absorbed by her, becoming little more than a clump of sperm-producing cells. Spoon worms can boast the animal kingdom's greatest disparity in male and female body sizes. In one species, an 8 cm female can contain a male often only 1-3 millimetres long. In other species of spoon worms, the males have never been found. By contrast, most angelfish and butterflyfish lead boringly conventional sex lives. Both sexes look alike; most mate for life, and are rarely seen more than a few meters away from their mate.

Some sea creatures maintain conventional sexual identities, but reverse the normal parental roles. In most damselfish, jawfish, cardinalfish, and seahorses, the eggs are brooded by the male (in his mouth, in the case of jawfish and cardinalfish). In one species of frogfish the female lays eggs on the side of the male's body, where he cares for them until they hatch. Male seahorses have a ventral pouch where the female deposits the eggs for brooding. They appear to actually go through labour pains while "giving birth" after the eggs hatch. A different type of role reversal occurs in one type of mantis shrimp, where the females aggressively pursue the males for mating, sometimes injuring them as they strike them with their claws.

Some brine shrimp have dispensed with males altogether, and form "Amazon" populations which are all female. Sponge shrimp colonies are mostly female, but, like bees and ants, produce a few drone males to fertilise the queen's eggs. As in social insects, the queen shrimp is the only one which reproduces. The worker castes are all sterile females.

DOUG PERRINE

As do birds, humans, and other terrestrial life forms, many sea creatures help to ensure successful mating through elaborate behavioural rituals, costume changes, and communication by visual, audible, chemical, and possibly even electrical signals. It is mesmerising to watch the neon-like colour patterns shifting over the bodies of unicornfish, squid, or cuttlefish during courtship. The foghorn-like booming of toadfish trying to attract a mate and the sequences of bizarre noises emitted by male humpbacks in breeding season can be deafening underwater, even at a distance of a hundred meters. Small bottom-dwelling reef fish often swim ritualised "dance" patterns and/or flash bright patterns hidden under their fins. Stone crabs scratch the striated "fingerprints" on their claws to "telegraph" the opposite sex, while fiddler crabs raise up a brightly coloured claw and waive it around. Flashlight fish, like fireflies, blink lights to coordinate their nighttime reproductive activities. Deep-sea fishes likely use bioluminescence for sexual communication as well as other activities. Some skates can apparently signal each other in the dark for sexual purposes using electrical impulses. Females of many species release pheromones that attract males, which may have to overcome fierce competition to earn the opportunity to mate. Whales, dolphins, manatees, turtles, rays, and sharks are among those where a female in heat can attract a "rowdy group" of males, which may seriously injure each other while competing for access to the female.

A pair of Orange-lined cardinalfish *Apogon cyanosoma*. The female extrudes eggs which are fertilised by the male. Then the male (left) takes them into his mouth for brooding. Photos from Mabul Island, Malaysia.

Reef squid are more sensible and compete for mates with colour displays. A male will often display one pattern on the side of his body facing towards the female he is courting, and another pattern on the opposite side, directed toward a competing male. Male right whales do not compete for mates by fighting, like sperm whales and humpbacks, but rather compete to produce the most semen - the better to wash out the seed of the female's last suitor and replace it with their own. This "arms race" has produced the largest gonads in the animal kingdom. The combined weight of the testes of a mature right whale can be over a ton. The penis is also a natural wonder, perhaps the longest in our solar system, but relatively narrow and flexible. In action it looks sort of like an aircraft refuelling hose with an internal guidance system. That it is able to locate and connect with a moving, and often resisting, target that is out of sight and 10m or so away from its head is quite remarkable. Successful copulation often depends on the presence of other suitors who prevent the female from escaping. These often take a turn afterwards, earning right whales the moniker of the "frat boys of the sea."

Most invertebrates are spawners, but some practice crude forms of internal fertilisation. Cephalopods use an elongated arm or tentacle to slip a sperm packet inside the female's mantle. (You might say that they like to practice sex at arm's length - probably a wise strategy for the male, given the female's predilection for cannibalism in some species.) Some barnacles use a penis which is the longest in the animal kingdom, in proportion to body size, to squirt sperm into their neighbour's shell (but they can't crow too loud about their masculine prowess because they are female at the same time).

By contrast, some corals are "all male"(each polyp in the colony is male, while all the polyps in other colonies are female). However, they don't crow too loud either, as they have no penis at all. A barnacle-type organ would-n't do them much good anyway, since their neighbours are all male too, not to mention exact clones of themselves. (Even if there was a female next door, what's the chance she'd want to do it with a guy that lives with his parents, has no job, no car, and his bottom glued to the couch, eats fish feces for breakfast, and has his anus on his face?) So they do the sensible thing and spit their sperm out their mouths (we call it "smoking") where it can drift away in the current and fertilise the eggs of some lucky females who've never met them.

In other species of coral, all of the polyps in a colony are both male and female. At spawning time, they spit out a bundle that looks like a big egg, but is actually a bunch of small eggs and sperm all packaged together. The package opens at the surface, and the various gametes go looking for partners, but miraculously avoid fertilising each other. Releasing eggs to the currents is called "broadcast spawning." Some types of corals and sponges are broadcast spawners, but others are brooders, with the males releasing sperm into the water, and the females drawing it into their body cavities to fertilise their eggs. Broadcast-spawning corals and sponges need to coordinate their releases. It is believed that the actual release occurs as a chain reaction, triggered by chemical signals from the first individuals to go off, but that the preparation is prompted by a variety of cues, including lunar phase, water temperature, tidal level, day length, etc. Whatever the actual signals that set it off, the underwater blizzard released during a mass spawning is a spectacular sight. In some areas, multiple species coordinate their spawnings on the same night, overwhelming the ability of egg-eating predators to consume the vast amounts of material released.

RON THRESHER

The courtship of the Emperor angelfish *Pomacanthus imperator*. The larger and more brilliantly coloured male at first swims below the female, then both ascent in the water column to spawn. Photos from Taiwan.

While sexual reproduction is dominant among most creatures large enough to be visible to the naked eye, asexual reproduction is also important. Fragmentation - reproducing from broken bits - is an important method by which corals, sponges and other creatures can recover from a disaster and repopulate an area after a major storm. Some corals and anemones can clone themselves by budding and/or fission. Some sea stars can reproduce either sexually (by spawning), by fission (splitting down the middle), or by autotomy (casting off a part which then regenerates the rest of the animal). Some species can grow an entire body from a single cast-off arm. Sponges are the regeneration champions, though. In lab experiments sponges have been ground up in a blender, then strained through a cheese cloth to separate the cells, and the cells were able to reform themselves into a new sponge!

In the sea, sex is all around you all the time. Dawn and dusk dives are generally preferred by subsea voyeurs due to the large numbers of reef fish that spawn at these times. However, it is a rare dive where some type of reproductive activity is not visible to a keen observer. Even clear seawater contains substantial portions of gametes and larvae, as well as water (much to the dismay of those who have personally discovered the connection between the spawning of the sea thimble jelly and the malady known as "sea bather's eruption.") It's no wonder that many divers describe their underwater experiences as highly sensual - they are actually swimming in a sexual soup.

ANCHOVIES ENGRAULIDAE

Thryssa baelama **Flores, Indonesia**

Little priest
Length: To 15 cm.
Distribution: Indo-Pacific from Red Sea to Samoa, including all of our area. Depth: 1-200 m.
General: Forms large and dense schools in coastal bays, often congregating near fresh-water run-offs. Fast-swimming fishes, darting about in feeding frenzies when encountering a cloud of plankton.
Below: **Goldspot herring** *Herklotsichthys quadrimaculatus,* to 14 cm, entire area, 1-40 m. **Herrings** (family Clupeidae) are a large family with 65 genera and 180 spp. world-wide, constituting an important food source for predators and man. Divers cannot readily identify anchovy and herring species.

SILVERSIDES ATHERINIDAE

Atherinomorus endrachtensis **Halmahera, Indonesia**

Striped silverside

Length: To 10 cm.
Distribution: Philippines and Palau to PNG.
Depth: 1-15 m.
General: During the day, this species aggregates in huge inactive schools, which stay close to calm shores. Feeding takes place mostly at night, mainly on zooplankton. Upper edge of silvery stripe along the side of the body with iridescent blue line. Pectoral fin area without dark spot. For a positive identification one has to hold the fish in the hand and open its mouth to check lower jaw elevation.

Most family members are small, slender, zooplankton eaters schooling in marine or estuarine waters. They are important prey for other fishes, e.g. jacks and even sharks. Schooling behaviour offers a certain degree of shelter for the individual as even fast and efficient predators have difficulty focusing on one.

MILKFISHES — CHANIDAE

Milkfish

Length: To 180 cm.
Distribution: All of our area.
Depth: 0 - 30 m.
General: The streamlined milkfish occurs solitarily or in small schools in a wide variety of habitats ranging from fresh- to saltwater, from mangroves to coral reef lagoons and outer reef slopes. It feeds on benthic algae and sieves sediment for invertebrate fauna. It is also often encountered with mouth open feeding on the algal scum directly at the surface of the water. The family is monotypic which means it comprises only one species.

Chanos chanos **Mabul I., Malaysia**

NEEDLEFISHES — BELONIDAE

Crocodile needlefish

Le: To 130 cm. Di: Globally in tropical waters, all of our area. De: 0-5 m. Ge: The largest family member. Pelagic but often patrols over the top of reef margins, including coastal zones. Several similar but slightly more slender species occur near mangroves.

Needlefishes have extremely elongated jaws with numerous needle-like teeth. They are potentially dangerous when attracted by light at night, there are known fatalities of leaping fish killing fishermen. Some species in freshwater.

Tylosurus crocodilus **Flores, Indonesia**

HALFBEAKS — HEMIRAMPHIDAE

Barred halfbeak

Le: To 35 cm. Di: All of our area. De: 0-5 m. Ge: Coastal bays and lagoons over shallow sand or reef flats, usually in the vicinity of seagrass beds, swimming in small loose groups. Juveniles are barred, large adults have a series of spots along back. Surface fishes with elongated to greatly extended lower jaw in most School in coastal and estuarine waters, some in freshwater. Family members mostly shiny and silvery, feed on a variety of algae and zooplankton. Sticky filaments on eggs, also viviparous.

Hemiramphus far **Kerama I., Japan**

Plotosus lineatus **Talaud I., Indonesia**

Plotosus lineatus **Cebu, Philippines**

Plotosus lineatus **Banda Sea, Indonesia**

Striped eel catfish

Le: To 35 cm. Di: Indo-Pacific, including all of our area. De: 1-50+ m. Ge: Most of the world's 2,000 catfish species occur in freshwater. Only 2 families with a small number of species are marine. The Striped eel catfish is the only common one to be encountered on coral reefs in our area. Also found along open coasts, in estuaries, and in tidal pools. Juveniles (centre photo) are commonly seen out in the open, feeding on mud, sandy slopes or seagrass beds in dense, ball-shaped groups. The smaller the juvenile the denser the schooling. They live mostly in coastal estuarine habitats with terrestrial debris such as sticks, coconuts etc. Large adults (bottom photo, compare to adult snappers) are solitary and hide under ledges during the day. Their major food items are small crustaceans, molluscs and fishes.

There are two main types of catfish, the most obvious difference is seen in the shape of the tail. Members of the family Plotosidae have an eel-like tail with dorsal, anal and caudal fin connected into one confluent fin. Members of other families have a separate forked caudal.

Avoid handling these fishes as their fin spines are highly venomous. Stings are extremely painful and repeated stings can be fatal. If stung apply heat (water or air-blower as hot as one can bare) to the wound as soon as possible to degrade the protein-type venom. After such first aid seek professional medical assistance.

CONVICT BLENNIES PHOLIDICHTHYIDAE

Convict blenny
Length: To 30 cm. Distribution: Indonesia to Philippines and Papua New Guinea. Depth: 1-50 m. General: This distinct little fish is found from sheltered coastal bays to deep outer reef walls. Only small juveniles are commonly observed. They are seen nearly always in massive schools (top right and small photo below) which spread out in the open water when feeding on zooplankton over a great distance or depth. They swim closely packed when on the move or threatened. Their coloration resembles that of the young of the venomous Striped eel catfish (see previous page) but they don't have barbels in the mouth area. At a length of about 10 cm the striped pattern begins to break up into spots and the fish settle in pairs on the substrate. Adults (centre right photo) are spotted and only rarely seen.

This family is systematically not placed correctly here but in this popular identification book the possibility of direct comparison with the Striped eel catfish (mimic model for juveniles) is preferred.

Pholidichthys leucotaenia **Alor I., Indonesia**

Pholidichthys leucotaenia **Batangas, Philippines**

BROTULAS OPHIDIIDAE

Bearded rockling
Le: To 60 cm. Di: Entire area. De: 3-50+ m. Ge: Coastal to protected inner reefs, rocks and coral with lots of crevices. Hunts on the reef at night (below) but quickly retreats when caught in a light beam. Eats fishes and crustaceans.

Brotula multibarbata **Flores, Indonesia**

A primarily tropical family with 2 genera and about 35 species worldwide. Most are reef predators, featuring a torpedo-shaped body, a large mouth with long needle-like teeth to seize prey. In tropical reefs they typically rest on top of the corals to look for possible prey. Camouflaged and waiting motionlessly in ambush, they take careless fish lightning-fast which are usually juveniles of larger reef dwellers. Most species are found on or near reefs, some on hard substrate but usually on sand patches on reef crests or rubble slopes, sometimes burying in the substrate. They can be found on coastal as well as outer reefs, depending on species. Identification of some species is difficult with only a few distinct features available and great similarities in general appearance. In addition there is strong variation within each species as is the case with many camouflaged species, matching surroundings or a particular habitat.

Lizardfishes produce small pelagic eggs (1 mm in diameter), hatchlings are about 3, postlarvae about 40-50 mm long. Juveniles bury in fine sand or mud adjacent to reef margins. Adults often form pairs, sometimes also small groups.

Synodus rubromarmoratus **Mabul I., Malaysia**

Red-marbled lizardfish
Le: To 12 cm. Di: Entire area. De: 10-50 m. Ge: Small, easily overlooked. Prefers clear protected reefs, usually along base of drop-offs on rock, coral. Banding more defined, snout pointed compared to similar spp. Below: **Grey-streak lizardfish** *S. dermatogenys*, to 22 cm, entire area, 1-50 m.

Synodus variegatus **Cebu, Philippines**

Variegated lizardfish
Le: To 25 cm. Di: Entire area. De: 3-50 m. Ge: Most common lizardfish on coastal to outer reefs. On hard substrate, including live coral and sponges, singly or in pairs. Colour variable, dull grey to bright red bands. Below: **Tail-blotch lizardfish** *S. jaculum*, 20 cm, entire area, 3-50+ m.

Painted lizardfish

Length: To 30 cm.
Distribution: Indo-Pacific, including all of our area (also found in the Atlantic Ocean).
Depth: 0.5-200+ m.
General: Inhabits protected coastal and estuarine sand flats from the subtidal zone to deep offshore habitats. Lives in groups spread out on sand flats and slopes and remains totally buried with only the eyes exposed. Only when chasing each other or when disturbed they come out of the sand, quickly dashing away to bury again as soon as possible.

Trachinocephalus myops **Kerama I., Japan**

Slender saury

Le: To 28 cm. Di: Widespread, entire area. De: 3-100+ m. Ge: Common on reefs, also in coastal outer lagoons and deep offshore. Highly variable, usually with 2 black saddles on back below dorsal and adipose fins. Below: **Blotched saury** *S. nebulosa,* to 16 cm, entire area, 0.5-100+ m, in mangroves.

Saurida gracilis **Flores, Indonesia**

Long grinner

Length: To 25 cm.
Distribution: Probably more widespread, but only known from southern Japan, Philippines and Indonesia. Depth: 10-100+ m.
General: Found on sandy slopes in shallow coastal but also deep lagoons, where it lies almost always buried in the substrate during the day. Comes out at night, to probably hunt nocturnally active cardinalfishes (apogonids) which swim close to the bottom in the habitats preferred by this species.

Saurida elongata **Bali, Indonesia**

Anomalops katoptron **Bali, Indonesia**

Anomalops katoptron **Sangihe I., Indonesia**

Two-fin flashlightfish

Le: To 28 cm.
Di: Indonesia, Malaysia, Philippines to southern Japan and Micronesia.
Depth: 20-400 m.
General: A deepwater species which occasionally is seen at night in groups at depths of about 20 m or in rather deep caves. It is easily recognised by two dorsal fins. The light organ below the eye is rotated to turn it on or off. The juvenile in the centre photo was found at a depth of 20 m on a very dark night, swimming solitarily about 1 m above the substrate, flashing its lights regularly in about 1-sec-intervals.

This is a small but interesting family with only 2 genera and 5 species all of which possess a luminous organ just below the eye. This organ is filled with bacteria that produce very bright light, enough to enable the fish to find and recognise prey when feeding on zooplankton at night. The fish can turn the light organ off or flash it on and off by rotating it or drawing a black flap of skin over it. Such blinking also may serve communication within a group. During the day these fishes hide in reefs and usually in deep water. At night they come out and rise to shallower depths where plankton is, especially in very dark nights without moon or with thick cloud cover.

Photoblepharon palpebratus **Bali, Indonesia**

One-fin flashlightfish

Le: To 11 cm. Di: Indonesia, Malaysia, Philippines to central Pacific. De: 10-50+ m. Ge: Coastal reefs with deep drop-offs, usually in large groups feeding just off the wall in the current at night. The single dorsal fin readily identifies this species. The light-organ can be covered by a flap of skin. Almost never seen during the day, except in very dark parts of ship-wrecks or caves. Rises from deep water on dark nights to shallow depths. A small species which is easily maintained in aquaria, however, it is extremely sensitive to light level changes or bright light.

This family contains the shallow water members of a highly specialised group which is unique in having a peculiar luring apparatus above the mouth, usually a modified first dorsal fin spine. This family comprises 12 genera and at least 41 species worldwide. They are mostly cryptic reef dwellers which attract prey with their lure which is specialised in each species by mimicking the food-source of its prey. A potential victim may see its favourite food, which in reality is a copy in shape and even motion, presented as a hard-to-resist easy meal - only to become a meal itself. The luring apparatus comes in two parts, the illicium or stalk which varies greatly in length between species, and the esca with is the actual luring bit representing various types of crustaceans, worms or even fish. The lure is waved and moved to copy the specific motion of its model.

Anglerfishes don't just position themselves anywhere but choose a spot where prey is likely to come along. The tropical species primarily feed on other fishes and with their expandable stomach can accomodate any large prey which passes through the mouth, including species longer then themselves. Many of the tropical species produce mucus egg rafts with numerous tiny eggs, but there are some which have demersal eggs and these are guarded by the female. Such eggs are much larger, up to 5 mm in diameter, and much less numerous, some 5,000 were counted in one species. Pelagic juveniles settle at a length of about 25 mm but demersal egg hatchlings settle on the substrate already at a length of about 10 mm.

These interesting fishes are easily kept in aquaria but introduction to community tanks is not recommended as the result is just <u>one</u> well-fed anglerfish. They need separate quarters and can be trained to take frozen food.

Shaggy anglerfish

Length: To 20 cm.
Distribution: Entire area, especially abundant in Indonesia and Malaysia.
Depth: 3-90 m.
General: Found in coastal bays with rich sponge growth, with leaf litter on sandy bottom, where they are well hidden by excellent camouflage and usually rest in shaded areas. Variable from beige, light yellow, orange, yellow-brown to black. Usually with darker streaks or elongate blotches. The unusual photo at the right nicely shows the extraordinarily large esca, the lure at the top end of the illicium, of this species.

Antennarius hispidus **Bali, Indonesia**

Antennarius nummifer **Singapore**

Spotfin anglerfish

Length: To 13 cm.
Distribution: Entire area.
Depth: 3-75 m.
General: This anglerfish species is primarily found in the brackish water of estuaries and river mouths, but also in reefs. Illicium of this species - flicked enticingly to attract fish prey - about as long as the second dorsal fin spine. Esca large and variable in form and colour, often resembling a small shrimp. Overall coloration very variable, often with a large ocellated dark spot at the base of the soft dorsal fin.

Antennarius striatus **Pulau Seribu, Indonesia**

Striped anglerfish

Length: To 22 cm.
Distribution: Entire area.
Depth: 20-50 m.
General: A beautiful and very variable frogfish species that usually has distinct stripes although it sometimes may also be uniformly black or blotched. This contrasts with the usually seen frogfish markings of circles that imitate sponges. Some specimens have long hair-like skin growths. Usually found deeper among debris or algae down sandy slopes. The photo nicely shows the esca or lure of this species.

Antennarius commersonii **Mabul I., Malaysia**

Giant anglerfish

Length: To 30 cm.
Distribution: Widespread in tropical Indo-Pacific, entire area.
Depth: 1-50 m.
General: Shallow coastal to deep habitats but nearly always associated with sponges. Commonly found on jetty pylons which typically are situated in protected parts of islands. Comes in almost every colour, there are, however, no reports of blue specimens. Largest family member in our area.

Clown anglerfish

Length: To 10 cm.
Distribution: Indo-Pacific, including all of our area.
Depth: 1-15 m.
Ge: Small juveniles are not cryptic but move about out in the open. They are very obvious, however, they do not resemble fishes but rather obviously poisonous nudibranchs. Adults grow to become more camouflaged by their colour pattern and thus attain a more cryptic life-style. Inhabit coastal reefs in rocky areas. As is typical for most family members, also this species is highly variable in colour but juveniles are usually bright yellow to white with orange to red markings.

Antennarius maculatus **Derawan I., Indonesia**

Painted anglerfish

Le: To 16 cm. Di: Indo-Pacific, including all of our area but with several regionally different forms. De: 3-50 m. Ge: A very common but also one of the most variable species, virtually coming in all colours except blue and green, matching most sponge colours. Difficult to identify, with local colour forms and probably subspecies. Found in coastal reefs, lagoons and on slopes with corals, sponges or at night just out in the open on sand or mud. Note esca on photo. Below: **Sargassum anglerfish** *Histrio histrio,* to 15 cm, entire area, at surface in floating weeds (sargassum). Jumps on top of weed raft to avoid predators (and photographers) from below.

Antennarius pictus **Lembeh Strait, Indonesia**

Diademichthys lineatus **Tioman I., Malaysia**

Long-snout clingfish

Le: To 50 mm. Di: Entire area. De: 3-30 m. Ge: Swims in the open near corals and is usually present when there are long-spined sea urchins on the reef. In coastal reefs and protected inner reefs. Like this one, most family members are striped.

The large family comprises about 35 genera and over 100 tiny species, many of which are only a few centimetres long when fully grown. These fishes are particularly numerous in warm-temperate seas where corals are replaced by weeds. Several species have yet to be described.

Discotrema crinophila **Flores, Indonesia**

Crinoid clingfish

Le: To 2.5 cm. Di: Entire area. De: 3-15 m. Ge: Several clingfish species have clear-cut preferences for different hosts. This one is almost always found living together with crinoids. Below: A similar but yet undescribed clingfish species from our area.

Discotrema lineata **Hong Kong**

Two-stripe featherstar clingfish

Length: To 30 mm.
Distribution: Indo-Pacific, including all of our area.
Depth: 10-40 m.
General: Associates mainly with featherstars (crinoids), living amongst their tentacles but also often swimming around them. Occasionally found together with long-spined sea urchins. It is not as common as the previous species and differs by an extra white line on the sides and a more pointed snout. Also it seems to live deeper and to have a preference for dark hosts.

OBSERVATION ON JAWFISHES

The family Opistognathidae is quite speciose. Its members are variously known as jawfishes, missing links, pugs, harlequins, smilers, grinners or monkeyfish and are distributed in all oceans around the globe. As is the case with many popular names, the name jawfishes unmistakably points out the most obvious feature of these little troglodytes who live in self-made burrows: The large jaws which make up the huge mouth. Investing a lot of patience, Harald Januschke has observed the everyday life, habits and communication of the Goldspecs-jawfish from Asia Pacific.

PHOTOS: HARALD JANUSCHKE

A rare photo: a Goldspecs-jawfish out in the open.

The systematic position of the jawfish family has been a matter of dispute for a long time. Only when the brooding behaviour of jawfishes was studied their close relationship with dottybacks (Pseudochromidae) and basslets (Grammidae) became obvious. Shared characters are the disrupted lateral line, the ventral fins inserting at the level of the pectoral fins, the habit of building and living in burrows, and mouthbrooding. In contrast to the dottybacks, jawfishes love to inhabit sandy flats which do not provide any further shelter apart from their own burrows.

Most of the time only the heads of these troglodytes peek curiously from the holes in the sand, but with a little luck one can observe them while acquiring building material or while feeding. Then they dash from their burrow and snatch plankton that drifts past them. During these moments they are easy prey for vagrant predators. As jawfishes have no further means of defence their only hope is to return as fast as possible back into their burrows. For that evolution has optimally equipped them: The large head with laterally positioned eyes allows for a splendid 360° view. And the slender and - compared to the head - narrow, snake-like body allows for a lightning-fast retreat backwards into the burrow without losing visual contact with their surroundings. As they mainly live in colonies there also are uninhabited burrows which can readily be used for a quick retreat whenever the situation calls for one.

Most interesting is the brooding behaviour of jawfishes because they guard their offspring as so-called mouthbrooders. After the deposition of the eggs the male fertilises them in the burrow of the female and consequently takes them into his mouth. So far, nobody succeeded in documenting this event on film in the natural habitat of the jawfishes, but it has been witnessed in aquaria. During the mouthbrooding season one can observe the male with hundreds of eggs in his mouth at the entrance of the burrow, occupied with supplying the offspring with fresh water. In case of danger the eggs are spat out inside the burrow. After about 5 days the eggs hatch and the fry are on their own.

To find a jawfish outside its burrow one needs to be a little lucky and to photograph it in full size is a challenge for

A pair of jawfishes - or are they rivals? What happens next?

every underwater photographer. A 90-minute-dive in Lembeh Strait/Sulawesi was necessary to capture the full beauty of the Goldspecs-jawfish *Opistognathus randalli* on film. Or to put it in other words: Lying motionlessly in the sand and waiting for the fish to finally lose its fear; until hunger becomes overwhelming or a prey item passing by is so seducing that the fish leaves its burrow; until the fish forgets that there is somebody with a camera very close nearby who waits for it to come out. Often a shot was possible only after many long minutes, afterwards waiting for the wary sand-dweller continued. Absolutely nothing happens without self-imposed paralysis and actually one shouldn't breathe at all if one wants to photograph jawfishes. In such a way time seems to pass quite slowly until the whole film is exposed. Fortunately the depicted scenes have taken place in the shallow parts of the Kungunkan house reef and no care had to be taken about bottom time. Thus my pains have led to further interesting observations:

Often jawfishes live in small colonies where they build their homes, keeping an adequate distance to each other - my home is my castle. They find the necessary building material - small stones and pieces of dead coral - in the immediate vicinity of their burrows. Simplified acquisition of building material or - to put it in simpler words - robbing the neighbour spares lengthy searches and is usual practice among jawfishes. Who ever had a thief for a neighbour surely will keep a safe distance to his fellows in the future. Watching such a visit to the neighbours is a real thrill: According to the reaction of the Goldspecs-jawfish residing in its burrow it had not really invited its neighbour for a friendly visit. With its mouth wide agape it rather clearly signalled to the visitor that he was not at all welcome. Viewed through the camera seeker the gaping mouth was so impressive that even a fully-grown Swabian gets frightened. One can only speculate about the reason for the visit as the visitor went away soon after these threat displays. Perhaps they only served in territory demarcation or they even were the beginning of courtship between both sexes. As is known from the literature, it is only during courtship that the males of jawfishes become discernible from the females by assuming splendid courtship colorations. However, during this dive no further approaches were undertaken. Both opponents remained peacefully in their own burrows rolling their beautiful large eyes.

Presumably an unwanted neighbour, or else...?

Male jawfish carrying brood in his mouth. Mabul, Malaysia. Depth 22 m.

Their brooding habits indicate the systematic relationship of jawfishes with dottybacks.

A large, globally distributed tropical family with two distinct subfamilies: Holocentrinae with three genera, known as squirrelfishes, all of which are characterised by a prominent spine (thought to be venomous in some species) on the lower corner of the gill cover, and Myripristinae with five genera, known as soldierfishes, in which the spine on the gill cover is small. All in all there are about 70 species, most of which are red in colour. They are nocturnally active but also commonly seen during the day in caves or under large overhangs along drop-offs. Some occur solitarily and are secretive, others school in caves or near coral bommies. At night they hunt for shrimps or larger zooplankton and swim individually over reefs or adjacent sandy areas.

The smallest larvae known are about 2 mm long. The larvae of Holocentrinae have a long spiny rostrum at eye level which is lost or just evident in postlarvae. They settle when about 40-50 mm long but some pelagic specimens about 60 mm in length were photographed during a night dive off Sulawesi.

Giant squirrelfish

Le: To 45 cm. Di: Entire area. De: 6-100+ m. Ge: Shallow coastal to deep outer reefs. By day in caves, often in pairs, sometimes also in groups (photo). Largest squirrelfish and known by many common names like Spiny-, Sabre-, and Long-jawed squirrelfish.

Sargocentron spiniferum **Sumatra**

Redcoat squirrelfish

Le: To 27 cm. Di: Entire area. De: 6-50 m. Ge: Coastal reefs, large lagoons. In small to large groups on slopes and flats with large corals. Below: **Finelined squirrelfish** *S. microstoma,* to 20 cm, entire area, oceanic islands, 3-50 m. Rocky reefs, in our area mainly at the oceanic margins. Very secretive.

Sargocentron rubrum **Kerama I., Japan**

Sargocentron diadema **Singapore**

Crown squirrelfish
Le: To 17 cm. Di: Entire area. De: 2-30 m. Ge: Protected clear coastal and inner reefs along drop-offs and lagoons. Often seen by day singly or in small groups near coral heads, under ledges. Below: **Samurai squirrelfish** *S. ittodai,* 20 cm, entire area, 10-50 m. Outer lagoons, slopes. Often deep.

Sargocentron violaceum **Sangihe I., Indonesia**

Violet squirrelfish
Le: To 25 cm. Di: Entire area. De: 6-30 m. Ge: Clear coastal to protected inner reef slopes and walls with rich coral growth. Secretive by day. Adult with bright red head. Below: **Three-spot squirrelfish** *S. melanospilos,* to 27 cm, entire area, 3-50 m. Easily identified by 3 black spots near the tail.

Neoniphon aurolineatus **Okinawa, Japan**

Yellow-striped squirrelfish
Le: To 22 cm. Di: Indo-West Pacific, entire area. De: 40+ m. Ge: Clear, current-rich outer reefs. Rarely seen because usually below 40 m. Below: **Silver squirrelfish** *N. argenteus*, to 24 cm, entire area, 15-50 m. Coastal and protected inner reef slopes with rich coral growth and some current.

Blood-drop squirrelfish
Le: To 24 cm. Di: Entire area. De: 2-50 m. Ge: In reefs, sheltered lagoons, seagrass beds with coral patches, on deep slopes with rich coral growth. By day in groups, at night scattering over the reef to hunt crustaceans. Below: **Mouthfin squirrelfish** *N. opercularis,* to 35 cm, entire area, 6-50 m.

Neoniphon sammara Kalimantan

Splendid soldierfish
Le: To 25 cm. Di: Entire area. De: 15-50+ m. Ge: Uncommon, mainly oceanic or very deep. Clear deep lagoons, by day hovering close to their shelter. Singly or in pairs, however, schooling in some oceanic localities. Below: **Bronze soldierfish** *M. adusta,* to 35 cm, entire area, 6-50+ m.

Myripristis melanosticta Nabucco I., Indonesia

Crimson soldierfish

Length: To 25 cm. Distribution: Entire area. Depth: 6-50 m. General: Common but secretive during the day, often confused with other similar species of which there are several. Coastal, often found on silty rocky reefs and in shipwrecks.

Myripristis murdjan Cebu, Philippines

SOLDIERFISHES — HOLOCENTRIDAE

Myripistis chryseres Ogasawara Is., Japan

Yellowfin soldierfish

Le: To 20 cm. Di: Entire area. De: 12-200+ m. Ge: Mainly below 30 m depth, rarely in caves as shallow as 12 m. Below: **Violet soldierfish** *M. violacea*, to 20 cm, entire area, 10-50 m. Common, along deep drop-offs, by day in caves, forming small groups.

SEAMOTHS — PEGASIDAE

Eurypegasus draconis Singapore

Eurypegasus draconis Bali, Indonesia

Little seamoth

Length: To 8 cm.
Distribution: Indo-Pacific, including all of our area.
De: 3-15 m.
Ge: Coloration highly variable, matching surroundings, capable of quick changes. Mimicks bits of surrounding debris such as pieces of broken shell. Found in sheltered coastal bays on fine sand with sparse rubble. Extremely well camouflaged, partly buries when resting. Regularly discards outer layer of skin in one complete piece, apparently to prevent accumulation of epibiotic growths such as algae, hydroids and other organisms. Shown are a pair and a small juvenile.

Seamoths are a small family of small curious fishes, comprising 2 genera and 5 species. They are benthic, have horny depressed bodies and large pectoral fins kept horizontally, crawling on the bottom with ventral and pectoral fins. With long rostrum and inferior tube-like mouth, protrusible downward. Gill opening is a small hole facing upwards. Body encased in rigid bony plates, tail encircled by bony rings. Eggs and larvae are pelagic, small juveniles may be washed into tidal pools. The seamoth's diet consists primarily of tiny benthic crustaceans and other small invertebrates.

PEARLFISHES CARAPIDAE

Silver pearlfish

Length: To 17 cm.
Distribution: Entire area.
Depth: Depending on host.
General: Pearlfishes are only rarely seen out in the open because they have a commensal (or symbiotic?) relationship with sea cucumbers *(Bohadschia, Stichopus, Thelonota* spp.) and mussels. By day the translucent fish hide inside the gut cavity of their hosts which they enter and exit via mouth or anus with head or tail first. At night they leave their invertebrate partners to hunt for small prey.

Carapus homei **Palawan, Philippines**

TRUMPETFISHES AULOSTOMIDAE

Trumpetfish

Length: To 60 cm, doubtfully reported to 90 cm.
Distribution: Indo-Pacific, including all of our area.
Depth: 1-35 m.
Ge: Coloration of juvenile and adult different, also individuals can quickly change colour from longitudinal bands to vertical banding (see photos). A xanthic (yellow) form is common in some areas. Juveniles live among seagrass and in soft coral areas, adults from shallow reef flats to deep drop-offs, and from coastal waters to outer reefs, mostly solitarily, sometimes in loose pairs.

This family comprises only 1 genus and 2 (one Atlantic and one Indo-Pacific) species. Both are cunning predators, positioning themselves vertically along objects or riding closely over the back of other fishes to be able to approach prey (primarily other fishes) without being detected. *A. chinensis* is wide-ranging in tropical seas and expatriates to warm-temperate zones.

Aulostomus chinensis **Sipadan I., Malaysia**

Aulostomus chinensis **Java, Indonesia**

FLUTEMOUTHS FISTULARIIDAE

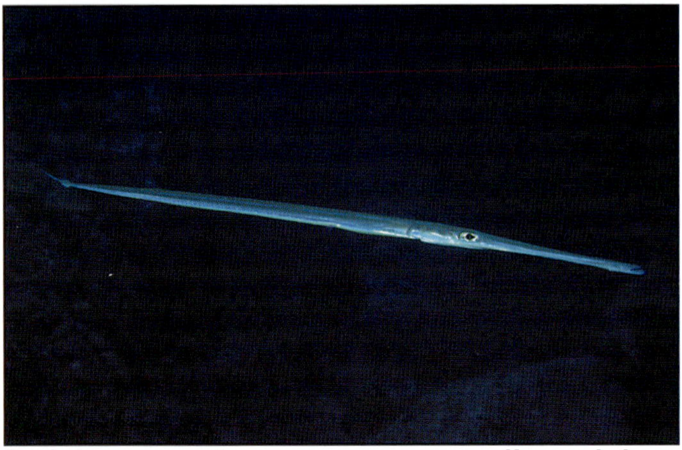

Fistularia commersonii **Kerama I., Japan**

Smooth flutemouth

Le: To 1.5 m. Di: Entire area. De: 1-100+ m. Ge: Adults dorsally with thin blue line or dashes. On coastal reef flats and in lagoons, often in small groups. At night with broadly banded pattern. See also centre photo on p.19, a flutemouth as prey of a snake eel.

SHRIMPFISHES CENTRISCIDAE

Aeoliscus strigatus **Palau**

Aeoliscus strigatus **Derawan I., Indonesia**

Coral shrimpfish
Le: To 14 cm. Di: Indo-Pacific, entire area. De: 1-30 m. Ge: On clear coastal reefs and in lagoons, schooling over branching corals. In lagoons often in small groups above long-spined sea urchins.

Below: **Rigid shrimpfish** *Centriscus scutatus,* to 15 cm, entire area, 3-100+ m. From coastal, often muddy to deep offshore habitats. Singly, in pairs or large dense schools. Individuals often seen sheltering next to sea pens or seagrass out in the open. Schools in rich coral patches with fan or whip corals.

Shrimpfishes are a small family of peculiar species which have adapted to a vertical position in life, swimming head-down. Pairs or large schools, comprising countless individuals, swim in a synchronised fashion, all turning at the same time. The two similar species common in our area are easily separated by a difference in their large dorsal fin spine which is rigid in *Centriscus* and hinged in *Aeoliscus.*

Solenostomus sp. **Tioman I., Malaysia**

Solenostomus sp. **Bali, Indonesia**

Solenostomus sp. **Irian Jaya**

Ghost pipefishes

A small tropical family closely related to Syngnathidae, the true pipefishes, comprising a single genus with a few described and apparently several clearly different but yet undescribed species (see photos left). They have short compressed bodies which are encased by bony plates, two separate dorsal fins, a large anal fin, similar and opposite to the 2nd dorsal fin, and a large ventral fin which in the female is enlarged and hooked onto the body to form a pouch for holding numerous small eggs (less then 1 mm in diameter). After about 3 weeks the tiny young hatch singly or a few at a time. It appears that they have a prolonged larval stage and settling juveniles are over half the adult size and ready to breed. Young adults are more slender than older specimens which develop a broader snout, caudal peduncle and body between dorsal and ventral fins. Such changes during growth have led to the description as different species.

Ghost pipefishes live exclusively in quiet environments, usually in coastal bays, hovering vertically with head down, near soft corals, fans on walls or among seagrass. They feed on shrimps and mysids, form pairs or small groups. Limited research suggests that males change into females.

Bottom photo: Juvenile overgrown with algae for camouflage. Below: *Solenostomus* sp. from Sulawesi, Indonesia. See also *Solenostomus* sp. on previous page, also from Sulawesi.

Robust ghost pipefish

Le: To 15 cm. Di: Entire area. De: 3-25 m. Ge: Colour variable from green to brownish-red, or blackish. Common in seagrass bed areas but very well camouflaged. Also on deeper coastal reefs in 15 to 25 m, coloration there usually reddish brown or blackish.

Solenostomus cyanopterus **Derawan I., Indonesia**

Ornate ghost pipefish

Le: To 10 cm. Di: Entire area. De: 3-25 m. Ge: Variable from red, white or yellow blotching (photo right), to almost totally black. On coastal reefs with rock faces or coral drop-offs, floating vertically close to the bottom. A common species, but easily overlooked.

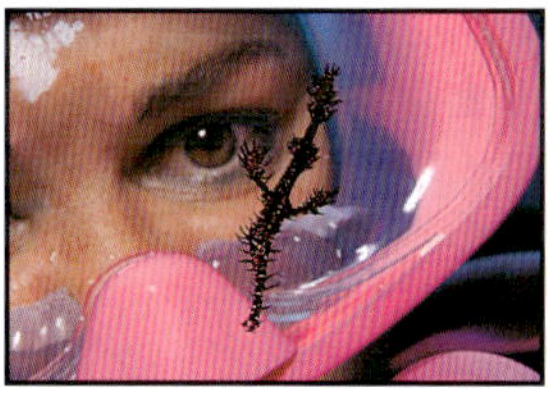

Solenostomus paradoxus **Ambon, Indonesia**

Delicate ghost pipefish

Le: To 10 cm. Di: Entire area. De: 15+ m. Ge: Coloration varies from brownish-red to pink. Adults often have an indistinct white band along the body. Found along reef edges close to sandy areas.

Below: Post-pelagical stage, length 8 cm.

Solenostomus leptosomus **Bali, Indonesia**

MOURNING FOR NAPOLEON

Sodium Cyanide? The environmentally conscious diver will surely recall that this is the chemical used by Philipino fishermen to capture tropical aquarium fish. The International Marine Alliance almost succeeded in bringing about a North American ban on the importation of tropical reef fish, and the use of sodium cyanide in the Philippines was banned only after the threat of such economic pressure. This was even a theme taken up for a German television documentary authored by Helmut Debelius in the eigthies. But this poison, also known as hydrocyanic acid, rears its ugly head again, this time a more widespread problem than ever. And all because a more lucrative business stands to gain from trade in coral reef dwellers.

PHOTOS: HARALD SCHMITT

Every diver's favourite: The Napoleon wrasse *Cheilinus undulatus*.

This new phenomenon is the result of an insatiable demand for living groupers and Napoleon wrasses (also called Giant humphead or Maori wrasse), the largest reef dwellers in the Indo-Pacific region. Prices are determined by strong demand among trendy restaurants in Hong Kong and Taiwan and more recently on the Chinese mainland. Rich Chinese are willing to pay top dollar (several thousand) to acquire living coral reef fish. The fish must be alive because it must be selected by the discerning gourmet before it winds up on his plate. The Chinese believe, you see, that fresh grouper and wrasses enhance virility. Does this sound familiar?

But what does sodium cyanide have to do with it? Sodium cyanide is used in strong concentrations as an underwater fish anesthetic. Indonesian cyanide divers call it obat or 'medicine'. Giant tablets are dissolved in large plastic bottles and sprayed at targeted fish hiding in the reef. Depending on the size of the fish, one bottle may be enough! The animal is quickly paralysed and can be picked up easily and brought to a boat on the surface. On board, the fish is placed into a seawater-filled container in which it slowly revives. It is then introduced into an underwater net cage where it awaits transfer to special Chinese ships. The living fish are taken aboard into holding tanks and transported to Hong Kong or Singapore. Thanks to the enormous demand, they may even be shipped via air!

In Asia coral reef fishes hiding in crevices and between corals are caught with the highly toxic sodium cyanide.

Just what is so objectionable about this trade? Groupers especially are a sought-after delicacy all over the world. It is the use of sodium cyanide which is objectionable. After all, fish with hook wounds are undesirable. It is quick and easy to catch fish with cyanide, but the dire consequence is the death of droves of invertebrates which cannot flee the water-borne poison. Years ago in the Philippines I determined the dose of cyanide appropriate for small fish. The disproportionately larger amount needed to anesthetise a full-grown five-foot grouper was apparent to me when I went diving in the cyanide-polluted reefs off Sumatra for the first time in 1995. I saw kilometres of desolate underwater desert, a shock-

When still alive, groupers, large parrotfishes and wrasses are highly sought after as food fishes for the customers of Chinese restaurants.

In large tanks the fish are transported alive to the market where they are bought by the restaurant kitchen managers.

ing spectacle. And at the same time it was a depressing sight because one could see that these reefs had once been alive. No storm or dynamite fishing could be responsible for such devastation.

But meanwhile, sodium cyanide fishing continues all over the Southeast Asian region. Currently the 17,000 islands of Indonesia are the major target areas for the capture of living Napoleon wrasses and groupers. In one way it is understandable, since a 17-year-old cyanide diver can earn $US 400 a month-three times what a university-educated civil servant makes. The local merchants who own the holding cages and organise the catch for the Hong Kong Chinese can be recognised easily in traffic; they drive the new BMWs seen cruising in a sea of Asian bonsai cars.

A living Napoleon wrasse is traded on the Hong Kong fish market.

Even Australia is not immune to Chinese greed. Asian boats filled with groupers have recently been intercepted off its northern shore. The first large study of the impact of the cyanide fishery appeared in October 1995. Its authors Johannes and Ripen gave the Indonesian reefs five years before the Chinese boats will have to move on to Papua New Guinea and Micronesia. "A tragedy with unforeseeable ecological consequences is unfolding here, and the world is taking no notice of it," complains Dr. Johannes of Tasmania.

In floating restaurants like this one the paying consumer can choose from a wide variety of live coral fishes.

This is not entirely true, since his study was quoted on the front pages of the daily press in the USA and Australia and was subject of some interest to weekly and monthly magazines in Germany, Holland and the UK. The fate of the Napoleon wrasse shows how quickly this plunder is impoverishing the seas. Years ago, Napoleon, famous for its tameness among divers from the Red Sea to Malaysia and to Palau, was available in every major Hong Kong restaurant as a prime delicacy. Today it has become a rarity and increasingly expensive. In ´jumbo`restaurants ships, one kilo costs $US 180 – and the price is rising. A dish of its lips, coveted as an aphrodisiac, fetches $US 300. In Hong Kong rich Chinese spend enormous amounts at wedding banquets or New Year´s celebrations. Thirty 12-seat tables, each provided with premium grouper and Napoleon wrasses, are not unusual. At 15,000 tons a year, Hong Kong is the largest market for reef fish in Southeast Asia. Taiwan (about 7,000 tons) and Singapore (1,000 tons) follow. Nearly two-thirds of all living coral reef fish offered in the restaurants of the region are captured with sodium cyanide: In 1995 this amounted to approximately 25,000 tons valued at one billion dollars. The following example shows that things are out of control: The Environmental Protection Minister of Indonesia, himself a diving enthusiast, banned the export of Napoleon wrasses. But this had the unintended consequence of making this variety of fish even more expensive. Currently the Chinese are willing to pay $US 100 per kilo of Napoleon wrasse, with individual fish weighing as much as twenty kilos!

Dr. Johannes sent me a video which graphically shows how the Napoleon wrasse and the surrounding reef suffer form cyanide divers. After being chased into a corner, the half-numb fish seeks shelter in the nooks and crannies of the coral reef and shoves itself into its deepest recesses. Two divers have to break up several metres of coral just to get at the grouper. Then, from a distance of a metre, a stronger shot of cyanide is aimed at its face. Now that the fish is totally defenseless, a line is run through the mouth and the gill opening of the pathetic victim. It is then pulled through the reef like a recalcitrant dog, trying to flee once again as it regains consciousness. In so doing it injures its head several times. Other effects of the cyanide are easy to see as a hundred small fish of various kinds lie dead on the sea floor.

Attempts to stop this practice have been fruitless. Cyanide tablets are, of course, officially banned all over Southeast Asia, but this ban is not enforced. In Indonesia, for example, it is illegal to fish with cyanide but not illegal to possess it. The Philippines are stricter: Instruments capable of detecting cyanide in fish tissues are available. With the support of the International Marine Alliance they are used on all live fish exports. But inspectors can be bought. In Irian Jaya the local Indonesians boarded the Chinese boats and killed the crews because of zero-yield fishing runs in their fishing grounds which had been polluted by Chinese sponsored cyanide fishing. The overall problem will not be solved in this way.

In the USA the danger posed by hydrocyanic acid to humans is well known as death sentences are carried out with it. Unfortunately, meat from cyanide captured fish is edible since the poison is concentrated not in its flesh but in its liver. A solution to this devastating problem in the Asia Pacific reefs is not yet in sight. The solution must be found in those countries which import large coral reef fish such as groupers, parrotfish and wrasses. Perhaps in such countries the following ‚substantiated' report should be issued as a press release to get the attention of the public: "The more cyanide-tainted fish is eaten, the faster the male genitals will shrink"!

<u>AUTHOR'S NOTE:</u> It is not my intention to offend anyone by publishing this documents; I only want to alert everyone to the long-term effects of such destructive practices. Dive tourism which is based on healthy reefs with a full spectrum of marine life will be far more rewarding on the long run to the local communities in the countries concerned.

The members of the family Syngnathidae are some of the most interesting fishes concerning both behaviour and appearance. This large family comprises over 50 genera and well over 200 species, distributed globally in all but the coldest seas and some occur in freshwater. There are two subfamilies, based on absence or presence of a caudal fin, including larval stages. Most pipefishes belong in the subfamily Syngnathinae, but some and the seahorses belong in the subfamily Hippocampinae. Tropical Seahorses are in need of revision as there are many similar species which are often not well known.

Syngnathids are mostly slender fishes, their bodies are encased in bony plates which are arranged in series of rings. All feature a small oblique mouth at the end of a tubular snout. The gills are enclosed within the gill chamber which is connected to the outside via a small pore-like opening above the opercle. Ventral fins and jaw teeth are absent. Anal and caudal fin may be very small or lacking in certain species or stages. The often long tail is prehensile in some species, especially the seahorses.

Usually the species live on shallow reefs or in seagrass beds, feeding primarily on small crustaceans which are sucked up with the long snout. They are distributed through a broad depth range, in protected bays some species occur as shallow as the intertidal zone, while off-shore some species are trawled in 400+ m. Their size range is also broad with some species maturing at about 25 mm in length and the largest species exceeds 65 cm. Their reproduction method is unique, as males incubate the eggs in a pouch or ventrally on the tail or belly, with or without a skin cover. Hatchlings are often well advanced and may already resemble their parents.

Hippocampus bargibanti **Lembeh Strait, Indonesia**

Pygmy seahorse
Le: 1-2 cm. Di: PNG to S-Japan. De: 25-50 m. Ge: Highly specialised species that has adapted to life on various soft corals on which is feeds. Part of the coral appears to live symbiotically in the skin and reacts into warty-like growth around the short spines on the head and body. This gives a perfect camouflage and matches colour and shape of its host in detail, and may seem like different species. Post-pelagic young settle on various hosts, but to breed they appear to prefer the red-polyped *Muricella* spp. that usually grow in depths over 20 m. Very difficult to see, a magnifying glass is useful. Far more common than originally thought it is only recently that this tiny sea horse has been recorded from many countries in the Indo-Pacific region. The original was described from specimens accidentally collected with a sea fan in New Caledonia. There appear to be varieties (species?) on different sea fans.
Facing page: Gulf of Thailand.

Hippocampus taeniopterus **Sulawesi**

Common seahorse

Length: To 22 cm.
Distribution: Moluccen Seas to Sulawesi and Bali, also to north and north-eastern Australia.
Depth: 5-15 m.
General: Females are sometimes yellow with several large dark spots on the trunk. Males are usually drab with striations over the head and small black spots over the trunk. Usually found along margins of seagrass beds, rarely deeper than 15 m. Adults occur usually in pairs. This species is generally referred to as *H. kuda.*

Hippocampus moluccensis **Lembeh Strait, Indonesia**

Moluccen seahorse

Le: To 16 cm. Di: Ambon and E-Sulawesi, possibly restricted to Moluccen Sea. De: 5-20 m. Ge: Common. Similar to previous sp. but coronet more distinct. On open sandy and muddy substrates from shallow protected bays on soft-bottom slopes to at least 20 m. Occasionally in pairs.

Hippocampus barbouri **Batangas, Philippines**

Zebra-snout seahorse

Length: To 15 cm.
Distribution: Only known from Philippines and northern Indonesia, but may occur in adjacent waters of Japan.
Depth: 6-10 m.
General: Previously confused with the closely related *H. histrix, H. angustus* and *H. spinossissimus.* Recognised by the spiny coronet, striped snout, double cheek-spines and thick spine in front of eyes on the snout. Unlike most seahorses, this species is often found clinging to hard corals and doesn't seem to be worried about stingers.

Right: **Estuary seahorse**
Le: To 15 cm. Di: Singapore to Java and possibly to Ryukyus, Japan. De: 2-16 m. Ge: Drab brown to black, often with filaments on crown and snout. Sub-marginal black line in dorsal fin. Females sometimes yellow with pupil-sized black spots on the trunk. Males usually drab with numerous small black spots over the body. Snout elongating with growth. In estuaries and harbours, entering brackish water. Adults in pairs. Most records of *H. kuda* are based on non-spiny species, especially the common seahorse *H. taeniopterus*.

Far right: **Great seahorse**
Le: To 25+ cm. Di: Tropical to sub-tropical Japan, probably China Sea. De: 20+ m. Ge: More slender than *H. kuda*. Trunk long, tip of snout expanded. Adults with distinctive fine white lines and spots in dark parts of head and trunk. Brownish to black, but juveniles and female sometimes cream or yellow. Mainly found on soft bottoms.

Hippocampus kuda **Bali**

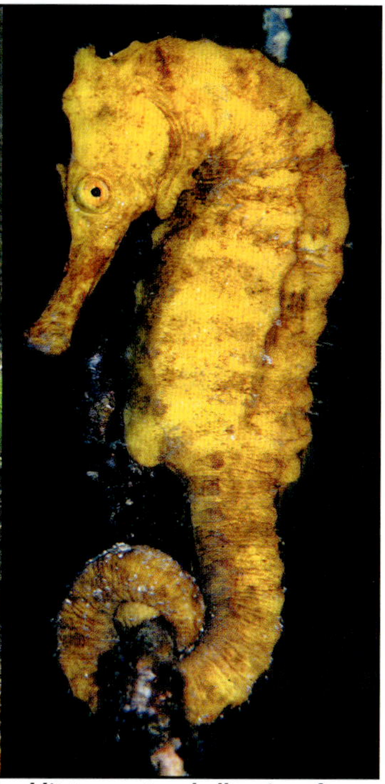

Hippocampus kelloggi **Japan**

Thorny seahorse
Le: To 15 cm. Di: West Pacific, mainly from Japan to Indonesia and Coral Sea. De: 10-15+ m. Ge: On soft bottom with soft corals and sponges, occasionally in shallower algae-rubble reef zones. Coloration highly variable to match surroundings. A distinct species with its spiny appearance and long snout, latter appears very long in juveniles and may elongate in large males. Sparse white barring on snout is diagnostic.

Hippocampus histrix **Lembeh Strait, Indonesia**

Doryrhamphus melanopleura **Bali, Indonesia**

Pacific blue-stripe pipefish
Le: To 6 cm. Di: West Pacific, Indonesia to southern Japan and Micronesia. De: 5-15 m. Ge: Blue stripe along upper sides is wide and lacks dark borders. Caudal fin variable and usually with elongated spot, centrally near its posterior margin. Males carry about 85 large eggs in a simple semi-open pouch under the trunk. Secretive under large coral pieces on shallow reef flats and lagoons, or in low caves where swimming upside down against the ceilings. Adults usually in pairs and pick parasites of other fishes.

Doryrhamphus japonicus **Iriomote I., Japan**

Honshu pipefish
Le: To 8.5 cm. Di: Subtropical Japan to Indonesia and Papua New Guinea. De: 0.5-25+ m. Ge: Three spots in the caudal fin and black-margined narrow blue stripe along upper sides of the body distinct. Easily confused with previous sp. that has a wider blue stripe and different caudal fin markings. On sheltered rocky reefs, usually in pairs near sponges or long-spined urchins. Active cleaner that shares crevices with shrimps, moray eels. From tide pools to at least 25 m, usually at moderate depths in subtropical zones, but in the tropics mostly shallow.

Doryrhamphus janssi **Mabul I., Malaysia**

Cleaner pipefish

Le: To 14 cm.
Distribution: Widespread in West Pacific.
Depth: 3-18 m.
General: This species has a much more elongated snout than other members of the genus. It is found on sheltered inner reefs, usually in caves with sponges and below large plate corals. This pipefish is a very active cleaner and has a cleaning station that is visited mainly by cardinalfishes and damselfishes where adults work in pairs. Usually swims upside-down.

Banded pipefish

Le: To 18 cm. Di: West Pacific. De: 1-10 m. Ge: Shallow water species, found inshore and in outer reef lagoons, sometimes in groups with numerous individuals, adults usually in pairs. Males carry eggs below trunk, hatch after a few weeks, young are almost transparent, and pelagic to about 30 mm. Pigmentation begins when settling on the substrate. Juveniles often near tidal zone, in rockpools. Often in large caves with shrimps or moray eels and engage in cleaning activities, probably picking small copepods or other parasitic crustacea from other fishes.

Dunckerocampus dactyliophorus **Nabucco I., Indonesia**

Many-bands pipefish

Le: To 19 cm. Di: Sumatra. De: 5-20 m. Ge: Sheltered reefs, usually in the back of caves, overhangs. Adults often in pairs. Nearly always swimming upside down against ceiling of caves. Below: **Yellow-banded Pipefish** *D. pessuliferus,* 16 cm, Philippines to Sulawesi. Deep mud bottoms, also in estuaries. Adults in pairs, active cleaner.

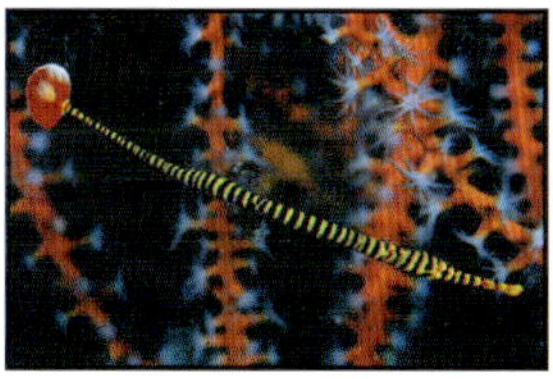

Dunckerocampus multiannulatus **Cebu, Philippines**

Bent stick pipefish

Le: To 40 cm.
Di: Widespread in Indo-West Pacific, but probably comprises several similar species that are distributed over the area. All populations need to be investigated further.
De: 2-25 m.
Ge: Populations in oceanic locations seem different from those in estuaries. Some populations inhabit seagrass beds and others only rubble sand areas. Most individuals seen on sand and mud areas, prone to currents. On soft bottoms the species is usually found to a depth of about 25 m.

Trachyrhamphus bicoarctatus **Bali, Indonesia**

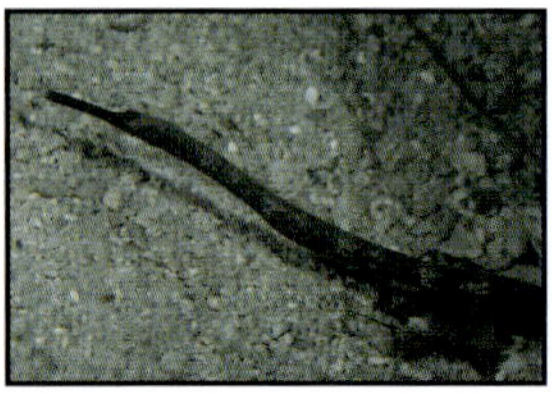

Trachyrhamphus longirostris **Bali, Indonesia**

Straight stick pipefish
Le: To 40 cm. Di: Tropical Indo-West Pacific. De: 3-50+ m. Ge: May be confused with *T. bicoarctatus* (previous sp.), but distinguished by lack of appendages on trunk in juveniles, head at no angle to the body, colour pattern. Less common, mainly known from deep trawls over mud, also in muddy estuaries.

Corythoichthys ocellatus **Bunaken I., Indonesia**

Orange-spotted pipefish

Length: To 12 cm.
Distribution: West Pacific, from Philippines to Solomon Islands.
Depth: 1-15 m.
General: The body is primarily covered with small to pupil-sized orange spots that are often dark-ringed as ocelli, especially over the back of the trunk. This small pipefish is found solitarily or in pairs, inshore, on algae and rubble reefs to a depth of about 15 m.

Siokunichthys nigrolineatus **Banda Sea, Indonesia**

Mushroom-coral pipefish
Le: To 8 cm. Di: Southern Indonesia and Papua New Guinea to Philippines. De: 10-20 m. Ge: Lives amongst the tentacles of mushroom corals such as *Heliofungia actiniformis*. Sometimes the pipefish is mistaken by divers for a worm and the coral for an anemone. The pipefishes occur in small groups, usually of mixed sex and sizes. Habitat at the bases of small drop-offs of inner reefs, often adjacent to large lagoons which discharge large volumes of water over reefs on outgoing tides that usually carry plankton.

DISCOVERY IN SECRET BAY

Many exciting discoveries in current marine biology, especially in the field of finding and describing new species (fishes and invertebrates alike) as well as finding out about their behaviour, have been made in coral reefs. This has been boosted by the development of modern diving gear, which makes this realm accessible to thousands of enthusiasts who go diving all over the world, albeit primarily in the warm and clear waters of tropical coral reefs. On the other hand, there are many different habitats which at first glance do not at all appeal to the devoted underwater researcher/photographer, but nonetheless harbour a wealth of lifeforms with yet unseen features and patterns of behaviour. Rudie Kuiter and Takamasa Tonozuka did not hesitate to enter the dark world of muck diving. Their recent findings on the muddy bottoms of Secret Bay are most astonishing!

PHOTOS: RUDIE KUITER

Diver picking up a holothurian or sea cucumber *(Bohadschia argus)* that immediately discharges its sticky white filaments in defence.

There is a tropical flatfish species that is commonly known as the Cockatoo flounder because of the numerous long filaments heading its dorsal fin. It is rarely seen by divers but primarily known from specimens that are trawled. It is apparently widespread in the West Pacific, ranging into the eastern Indian Ocean. The function of its peculiar filaments was never understood, but recent observations and photographs of a specimen (shown on these pages) from Gilimanuk, West Bali, in a shallow estuary that is now called Secret Bay, appropriately revealed this secret. *Samaris cristata* clearly shows a case of mimicry with sea cucumbers *(Bohadschia* spp.) that expel sticky white threads in defence when attacked or handled (see photos). The flounder, like most other species, simply relies on camouflage to avoid attention (see photo), but when accidentally discovered, touched or threatened it swings the anterior section of its dorsal fin suddenly forward to scare or deter potential trouble makers. The obvious white long rays appear and look typically like the sea cucumbers' sticky tentacles. An approaching predator that may have eaten a flounder before, is suddenly confronted with a troublesome sea cucumber and will quickly change its mind. The flounder looks very convincing, bearing the long and obvious white filaments that resemble the sticky tentacles of the sea cucumber, as well as the body showing a more striped pattern that is comparable with sea cucumbers of that size (the flounder is about 15 cm in

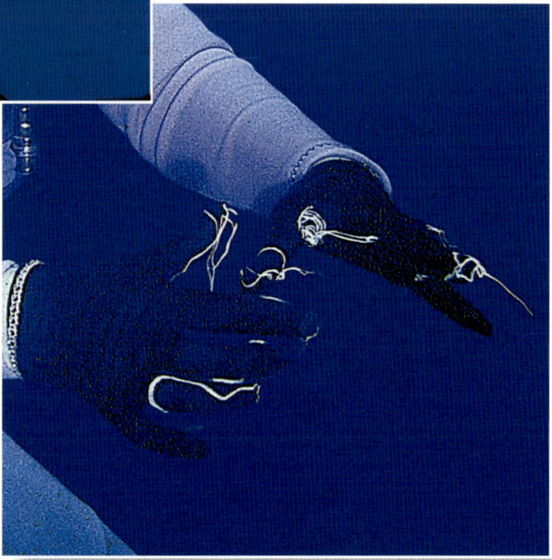

And they do stick! Nothing is safe of the sea cucumber's defensive appendages.

PHOTOS: TAKAMASA TONOZUKA

Cockatoo flounder, *Samaris cristata,* with its filamentous dorsal fin section spread out and thrusted forward to deter threatening predators (or bold underwater photographers).

total length, measured from tip of snout to end of tail). Flounders are best known for their ability to adapt their coloration to the immediate surroundings and in some cases juveniles are known to mimic flatworms or bits of shells. Some other flounders have features that have only recently been understood, such as the Angler flounder *(Asterorhombus intermedius)* which has its first dorsal ray modified like the antennariid anglerfishes into an esca and illicium, a lure to attract their fish prey. As divers are becoming more interested in habitats other than coral reefs, further amazing discoveries will undoubtedly be made amongst the flatfishes living on muddy substrates. Secret Bay in Bali is one of these places where many such discoveries are made on a regular basis by divers who are fascinated by mud critters.

This Cockatoo flounder in Secret Bay, Gilimanuk, Bali, Indonesia, has perfectly adapted to the ground (10 m).

A complex family with about ten subfamilies, two of which comprise most of the species: Pteroinae, the lionfishes, and Scorpaeninae, the true scorpionfishes. The lionfishes are the most spectacular family members because they have very long dorsal fin spines and large pectoral fins, often reaching past the anal fin. The scorpionfishes are a diverse group, comprising at least 12 genera which are widely distributed in all but the coldest seas. They have spiny heads and are mostly substrate hugging and very well camouflaged fishes. Despite their appearance most species are excellent food fish and in many areas some are of commercial value.

WARNING: All species possess venomous spines and a sting produces extreme pain, followed by numbness, and - although relatively rare - the stonefish of the subfamily Synanceiinae has caused fatalities. If stung, the best thing to do is to apply heat to the wound as soon as possible. Immersing the wound in very hot water is generally recommended but hot air from a hair-dryer or heater-radiator maybe more practicable, can be just as effective, and can be controlled better. Heat destroys the proteine-based venom when applied to the wound and its immediate surroundings, usually relief of pain is almost instantly.

Scorpionfish eggs and larvae are pelagic, generally very small, the eggs of about 1 mm in diameter are round or slightly ovoid, hatching larvae are about 2 mm long. Postlarvae can be less than 10 mm long depending on species, e.g. *Pterois* as small as 15 mm have been found on the substrate. All scorpaenids are predatory, some preying exclusively on other fishes but others may prefer crustaceans or a mixed variety of both. Whilst many species are very interesting or spectacularly looking and easily kept in an aquarium, it should be kept in mind that they generally have a large mouth and also a taste for expensive food species.

Common lionfish

Length: To 35 cm.
Distribution: Indo-West Pacific, throughout our area.
Depth: 1-50 m.
General: This species can be found almost anywhere on coral and rocks, from brackish estuaries to outer reefs. Its coloration is variable and has reddish-brown to black vertical bands. Some specimens found in estuaries are totally black, juveniles are also less colourful than adults and often grey with partly transparent fins. The sting of this large lionfish can cause extreme pain and severe cardiac problems, hence it is also known as devilfish or firefish. The voracious predator feeds primarily on fishes.

Pterois volitans **Palawan, Philippines**

Pterois antennata **Alor I., Indonesia**

Spotfin lionfish

Le: To 20 cm. Di: Entire area. De: 6-50 m. Ge: Brown bands of variable width, pectoral fin spines white. Coastal to offshore reefs from shallow flats to deep slopes. Eats fishes, crustaceans. Below: **White-lined lionfish** *P. radiata,* to 25 cm, body dark with white lines. Feeds on crustaceans.

Dendrochirus biocellatus **Batangas, Philippines**

Two-eyed lionfish

Length: To 20 cm.
Distribution: Indo-West Pacific, throughout our area.
Depth: 3-50 m.
General: A very distinct species with ocelli in the soft-rayed part of the dorsal fin and long barbels on the snout. Rarely seen during the day, but commonly seen at night in many areas on shallow coastal reef flats and in caves, like all lionfishes mainly nocturnally active. Usually seen singly but also pairs have been observed during courtship at dusk, males can be recognised by their grey (not black) ocelli.

Dendrochirus zebra **Nabucco I., Indonesia**

Zebra lionfish

Le: To 20 cm. Di: Entire area. De: 3-80 m. Ge: Coastal reefs and lagoons, shallow to deep water. The genus differs only slightly from *Pterois* by shorter fin rays and divided pectoral fin rays in adults. Below: **Dwarf lionfish** *D. brachypterus,* to 15 cm, entire area, 10-100 m. Coloration extremely variable.

Rhinopias aphanes **Komodo I., Indonesia**

Lacy scorpionfish
Le: To 25 cm. Di: West Pacific, eastern Papua New Guinea, Philippines to S-Japan.
De: 6-25 m. Ge: Rare but very distinct. On corals or sponges, extremely well camouflaged, obviously mimicking a crinoid. Variation below. See also previous page, photo also from Komodo Island, Indonesia.

Rhinopias frondosa **Sulawesi**

Weedy scorpionfish

Le: To 24 cm. Di: Entire area. De: 10-50+ m. Ge: Colours and body patterns very variable, markings if present tend to be nearly circular. Generally deeper than 10 m and in sheltered areas with slight currents not necessarily on coral reefs but also on soft bottoms.

Sebastapistes cyanostigma **Yonaguni I., Japan**

Yellow-spotted scorpionfish

Length: To 7 cm.
Distribution: Coral Sea to southern Japan.
Depth: 1-15 m.
General: A very distinctive small scorpionfish species, uniquely patterned with yellow blotches and white spots, found by searching among the branches of *Pocillopora* corals on reef tops of exposed coral reefs. A little known fish which is actually quite common.

Ambon scorpionfish
Le: To 8 cm. Di: Entire area. De: 30-50+ m. Ge: Usually in pairs on dark sand amongst debris. Horn-like growths above eyes and long mustaches distinct. Female larger, growth smooth (branching in male). Below: **Shortfin scorpionfish** *Scorpaenodes parvipinnis,* to 8.5 cm, 10-50 m. Often near sponges.

Pteroidichthys amboinensis **Sipadan I., Malaysia**

Demon stinger

Length: To 25 cm.
Distribution: Entire area.
Depth: 1-40 m.
General: A hazardous fish for divers since it often buries in sand during the day and can deliver a vicious sting. Like all family members active in the evening and at night. Colour is variable, some specimens are bright red. Identified from its pectoral fin pattern which is displayed when the fish is threatened. Walks rather than swims using two free pectoral fin rays and usually has its tail curled while doing so.

Inimicus didactylus **Sipadan I., Malaysia**

Leaf scorpionfish
Le: To 10 cm. Di: Entire area. De: 1-50+ m. Ge: Body unusually thin, leaf-like, white, black, yellow and violet. Often sitting on branches of staghorn or other exposed corals. Usually easy to photograph as it will stay in place. If one is found there are usually others nearby. Can be found during the day.

Taenianotus triacanthus **Banda Sea, Indonesia**

WATCH WHERE YOU STICK YOUR APPENDAGES!

Potentially lethal and almost invisible, the stonefish nestled beside the rock. With lufting motions of its large pectoral fins the fish scooped sand about its misshapen form, perfecting it's already perfect camouflage. It waited in ambush, motionless, watching the schools of reef fish hovering about its head. Clay Bryce tells about nasty incidents with the world's most poisonous fish.

PHOTOS: CLAY BRYCE

With a face only a mother could love, the stonefish is none the less a highly specialised fish that prefers to use its broken outline and natural camouflage abilities to stay hidden.

On the surface, between a cloudless blue sky and clear tropical waters, a group of excited divers completed their preparations. Like lemmings off a cliff, the aquanauts jumped from the boat into the water to sink in varying degrees of disorder. Totally pre-occupied with popping ears and a leaky facemask the first diver began his descent. He plummeted towards the ocean floor and the unsuspecting stonefish with all the grace and speed of a jettisoned weight belt. His ill-judged landing would prove to have that infallible flair for pin point accuracy that only the accident-prone possess!

With a survival strategy that relies on stillness and 13 potentially lethal spines the stonefish

With lifting motions of its pectoral fins the stonefish can cover itself with sand to enhance its natural camouflage abilities - stepping on one will definitely ruin your day!

Close-up of the dorsal spines of the stonefish. The venom is potentially lethal but secondary infection from the spine will cause a severe reaction.

A single spine with double venom sacs - a potent aid for a fish that prefers to lay still rather than swim from predators.

remained motionless beside its rock. As the descending diver plunged through the water-column the stonefish instinctively flared its dorsal fin and braced for impact. The sun darkened, the earth moved and the crystal clear waters around the fish filled with sediment.

The diver's left knee absorbed most of the force as it thudded into the substrate. His neoprene-encased right knee was not so lucky. The stonefish's sharp, hypodermic-like spines easily perforated the layer of wetsuit and taut skin below the right kneecap. As the spines penetrated further the venom glands were squeezed pumping toxin along the shafts of the spines into the soft flesh.

With surprising agility the diver jerked his right leg up and rolled on his back clutching the injured knee to his chest. He closed his eyes, his mind becoming fogged with the pain; a pain that steadily increased in intensity. The dive had started badly and was set to get a lot worse! The stonefish quietly moved to a less crowded rock.

Without immediate assistance our luckless diver would probably die from drowning as shock set in, or perhaps an embolism from a panic-induced rapid ascent. This fictional scenario could happen to any diver or reef walker as they explore the warm tropical waters of the Indo-Pacific. Even without the complication of being underwater, any unfortunate victim of a stonefish spiking can expect intense pain and distress that may last hours or even months. Ferocious pain, difficulty in breathing, delirium, seizures, vomiting, shock and even cardiac arrest are all to be expected as the myotoxic venom spreads. If not treated correctly, the tissue at the site of envenomation may become ulcerated, abscessed and gangrenous within a few days. Side effects can last several months and sensation at the site may never return.

Pterois volitans, like all lionfish has a painful venom associated with its spines - the stonefish venom is 10 times stronger!

The stonefish is probably the deadliest of all fishes. Venom toxicity is 10 times stronger than that of the lionfish, *Pterois volitans,* and the more spines involved the greater the effect! The famous ichthyologist, Dr. J.L.B. Smith, while working on an East African reef flat, describes his reactions to a stonefish sting in the scientific journal *Copeia,* 3 (August 31, 1951). His full account is hair-raising in the extreme and enough to keep you on dry land. Here is a summary of his ordeal, quoted from *Copeia,* beginning with the moment he was stung.

5-10 seconds "...intense stab of pain shot up my arm to my neck".

Less than 5 minutes "[pain]...of an intensity never before experienced".

After 5 minutes "...insane desire to ease the mounting agony by rolling on the ground".

Next 2 hours "...perspiration dripped to the floor,...little recollection of events, ...close to collapsing". (Local witch doctor unsuccessful!)

After 3.5 hours "...pain was undiminished and the perspiration still perfuse". IMMERSION IN HOT WATER FOR 4 HOURS. (Administered by his wife.)

7.5 hours "...intense agony no longer recurred on removal from hot water".

Next morning "...the thumb was greatly

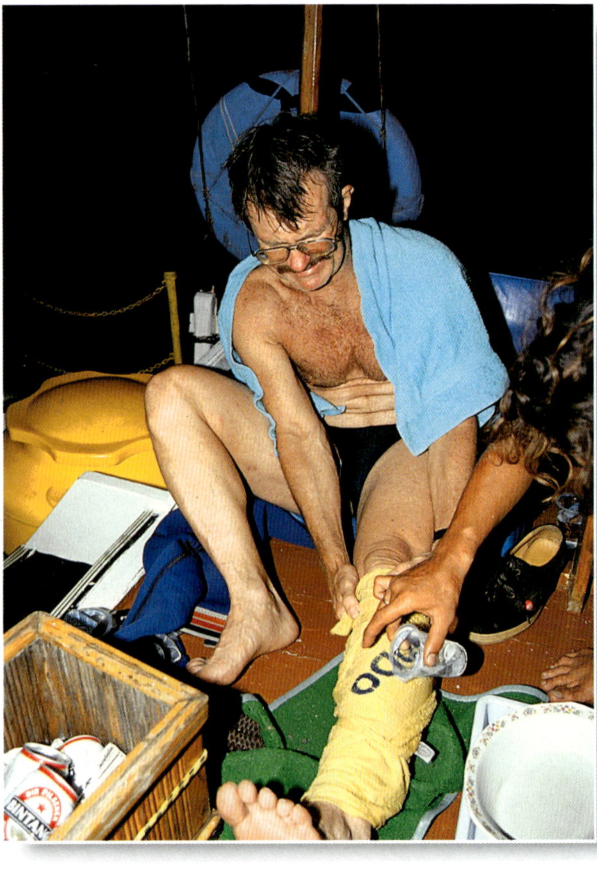

Dr. Gerry Allen, after a spiking to the leg from an angry lionfish. Like the treatment for a stonefish spiking, hot water, in this case poured over a towel to hold in the heat, is the best treatment. For stonefish, however, hospitalisation is highly recommended.

enlarged, ...turned black...., ...without sensation. Hand greatly swollen, also the lower forearm, ...intensely painful to touch".

24 hours "...large yellow blisters formed, ...that dripped steadily [when punctured] thereafter for 6 days".

3 days "...[swelling] reaching a maximum...".

6 days "...inflammation and pain in the thumb increased alarmingly and pus appeared. 1,000,000 units of penicillin in 5 doses over 16 hours - beneficial effect".

9 days "...some degree of sensation returned..., [swelling] no longer beyond the elbow".

30 days"...black portions commenced to fall away... The site of the cut opened to a discharging cavity and did not heal finally until 50 days after the stab".

80 days"...hand still weak, ...thumb barely moveable. ...toxin has had a marked adverse effect on my general health and condition".

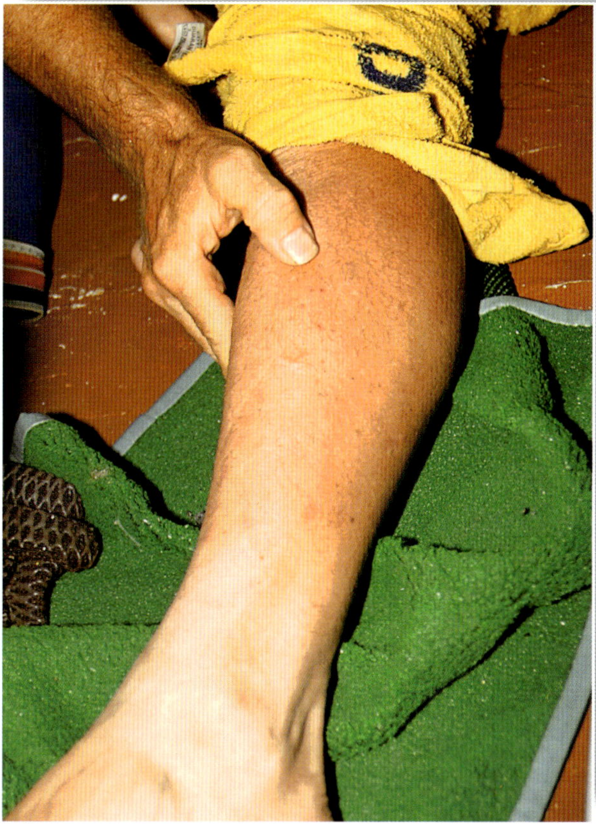

The small puncture marks (just below the thumb) from the lionfish spiking seem insignificant compared to the trouble they caused Dr. Allen.

In his article, Smith describes seeing many African natives who had been victims of stonefish stabs. They had missing toes and fingers which "would turn black and fall off" from gangrene due to secondary infection from the spine. The cool reasoning of Smith's wife saved him from losing his offended thumb through injections of penicillin; 1 million units of the antibiotic, which had to be repeated a week later! The use of hot water to denature the toxin and so render it ineffective, although belated, would have been a great relief to the agonized Smith. Hot water treatment for fish envenomation is now a standard first aid procedure thanks to the ingenuity of this lady.

The recommended first aid for stonefish envenomation is immediate intensive care at a hospital. However, given that divers like to explore out of the way places the distance to the nearest hospital is likely to take many painful hours or even days! In this case the immediate use of hot water is strongly recommended, as is the use of stonefish antivenene – something all of us carry wherever we go?! A hand or foot is easily immersed in hot water but a leg or upper arm is a little more difficult. Wrapping the limb in a towel soaked with hot water works well, with extra hot water being poured over the towel to maintain temperature. A little caution needs to be exercised over the temperature of the water; hot should read - as hot as the patient can stand - around 43 to 50°C. There is no need to add to the patient's woes by cooking them!

Dr. J.L B Smith - East African reef flat, Journal Copeia, No. 3, August 31, 1951

Next morning	"... the thumb was greatly enlarged,...turned black...., ...without sensation. Hand greatly swollen, also the lower forearm,...intensely painful to touch".
24 hours	".... large yellow blisters formed,.. that dripped steadily [when punctured] thereafter for 6 days."
6 days	"...inflammation and pain in the thumb increased alarmingly and pus appeared. 1,000,000 units of penicillin in 5 doses over 16 hours - beneficial effect."
30 days	"...black portions commenced to fall away.... The site of the cut opened to a discharging cavity and did not heal finally until 50 days after the stab."
80 days	"...hand still weak, ..thumb barely moveable. ....toxin has had a marked adverse effect on my general health and condition."

A summary from Dr. Smith's article about his spiking from a stonefish.

LEAF FISHES — TETRAROGIDAE

Ablabys macracanthus — Sulawesi

Spiny leaf fish
Le: To 18 cm. Di: Entire area. De: 10-50 m. Ge: Coastal sand and mud slopes adjacent to deep water. With its high-crested dorsal fin this species is very similar to the Indian Ocean *A. binotatus* but has more rays in the anal fin. Below: **Leaf fish** *A. taenianotus,* to 15 cm, entire area, 3-20 m.

FLYING GURNARDS — DACTYLOPTERIDAE

Dactyloptena orientalis — Palawan, Philippines

Flying gurnard

Length: To 32 cm.
Distribution: Very widespread, including all of the tropical Indo-Pacific, throughout our area.
Depth: 6-100+ m.
General: A spectacular species when displaying. Well camouflaged when resting and may partly bury in sand. Found in protected coastal bays on shallow sand flats and slopes with sparse reef or plant life, occasionally seen in pairs. Known also from trawls in excess of 100 m depth.

FLATHEADS — PLATYCEPHALIDAE

Thysanophrys otaisensis — Kerama I., Japan

Fringe-lip flathead

Length: To 25 cm.
Distribution: Widespread in tropical Indo-Pacific, including all of our area.
Depth: 1-30 m.
General: A common species in protected shallow sandy bays. Buries during the day and hunts at night, can then be seen sitting out in the open on the sand. There are several similar species, some with a longer snout, others differ in the spiny head ridges. Shown here is one of the species encountered most often.

The Serranidae are a highly diverse family with several major groups, but classification is continuously under review. Changes can be expected with regards to family status. As presently defined, there are nearly 50 genera and well over 400 species, distributed worldwide. They belong into 3 subfamilies: Epinephelinae, the groupers; Anthiinae, the basslets; Grammistinae, the soapfishes. Even these groups, especially the latter, are diverse assemblages of fishes, varying in size and shape or behaviour.

The Epinephelinae comprise the large groupers including the largest reef fish of all. Most of the larger species are of commercial importance, reaching 1-2 m in length. They feature tiny, often embedded scales and as adults have small egg-shaped eyes. Home-ranging on reefs, some species are vulnerable to spear-fishing, causing their decline in certain areas. Most groupers live solitarily and are territorial in caves or on larger sections of the reef. Some species may congregate in certain areas, at least annually, probably for mating. Their diet comprises a variety of fishes, cephalopods and crustaceans.

The Anthiinae or basslets are small tropical species most of which are schooling planktivores but a few live close to the substrate similar to the groupers. They feature small to moderate-sized ctenoid scales and large eyes, placed laterally and anteriorly on the head. The caudal fin is usually deeply emarginate to greatly lunate. Males of planktivorous species are often brightly coloured and look entirely different from females. In a swarm females greatly outnumber males and males derive via sex change from the larger, usually most dominant females which are in charge of groups of females in harem fashion.

The Grammistinae or soapfishes have a slimy skin which contains a bad-tasting toxin. They are the most diverse group within the Serranidae and several genera will eventually give rise to a new subfamily or even family. Presently these different assemblages within the subfamily are referred to as tribes. The soapfishes are usually more secretive than other serranids, living in caves or under ledges, often remaining close to or hidden in the substrate. Some species can be found in loose groups and show little fear towards divers.

Peacock grouper

Length: To 45 cm.
Distribution: Indo-Pacific, throughout our entire area.
Depth: 1-45 m.
General: Found from clear coastal reef crests to outer reef lagoons. Seen among rocks and corals, often slowly swimming about during the day. Variable in colour, the white conspicuous area around the pectoral fin base can be turned on and off.

Cephalopholis argus **Talaud I., Indonesia**

Cephalopholis sonnerati **Java, Indonesia**

Tomato grouper
Le: To 65 cm. Di: Indo-Pacific, throughout our area. De: 10-150+ m. Ge: Very small juveniles black with blue head and broad white margin on caudal fin, changing to brown and increasing orange spotting, adults bright red. Coastal slopes with rocks or coral. Adults deeper than juveniles.

Cephalopholis miniata **Pulau Seribu, Indonesia**

Jewel grouper
Le: To 40 cm. Di: Entire area. De: 3-50+ m. Ge: Clear coastal to outer reefs. Often in small groups. Along slopes and drop-offs, often in shipwrecks. Small juveniles orange, among coral. Below: **Flagtail grouper** *C. urodeta,* to 30 cm, entire area, 1-40 m, col. highly variable, Pacific form shown.

Cephalopholis sexmaculata **Mindoro, Philippines**

Saddle grouper
Le: To 45 cm. Di: Entire area. De: 0-150+ m. Ge: Clear coastal to outer reef drop-offs, usually in large caves with sponges, often upside down or vertical on walls. Feeds on apogonids, anthiines and shrimp. Below: **Orange grouper** *C. spiloparaea,* to 30 cm, entire area, 10-100+ m.

Bluelined grouper

Length: To 34 cm.
Distribution: West Pacific, throughout our area.
Depth: 2-20 m.
General: This grouper is a shallow water species most often encountered in sheltered silty reefs but also occurs in the vicinity of river mouths.

Cephalopholis formosa **Gulf of Thailand**

Harlequin grouper
Le: To 35 cm. Di: Entire area. De: 10-70+ m. Ge: Along clear outer reef drop-offs with large caves and ledges and lots of sponges. Remains under cover but can be closely approached. Juveniles are pink with fewer lines. Below. **Bluespotted grouper** *C. cyanostigma,* to 35 cm, entire area, 1-50 m.

Cephalopholis polleni **Cebu, Philippines**

Ozelot grouper
Le: To 85 cm. Di: Widely distributed, to Central America. De: 5-70 m. Ge: Lives in sheltered, rather turbid waters, but also found in clear water. Below: **Leopard grouper** *Cephalopholis leoparda,* to 25 cm, entire area, 3-40 m, secretive, one of the smallest groupers.

Dermatolepis striolatus **Maratua I., Indonesia**

Gracila albomarginata **Mindanao, Philippines**

Window grouper

Le: To 45 cm. Di: Indo-Pacific, throughout our area. De: 10-150+ m. Ge: Usually only seen on deep clear water drop-offs, in some areas where common it also occurs on shallow reefs. It swims well out in the open and juveniles (small photo below) resemble basslets.

Variola louti **Flores, Indonesia**

Moon grouper

Le: To 80 cm. Di: Indo-Pacific, throughout our area. De: 3-200+ m. Ge: Juvenile (below) with distinct colour pattern, changing gradually to all orange with small blue spots all over. On coastal shallow reef flats and outer reef crests and slopes. Juveniles secretive in rocky areas with boulders.

Variola albimarginata **Mabul I., Malaysia**

Lyretail grouper

Le: To 65 cm. Di: Entire area. De: 3-100+ m. Ge: Coastal to outer reef crests, usually in coral-rich areas. Caudal fin of adult with white posterior margin, juveniles orange. Below: **Barramundi** *Cromileptes altivelis*, to 65 cm, entire area, 1-40+ m. Juveniles distinctly spotted, genus monotypic.

Whitelined grouper
Le: To 52 cm. Di: Entire area. De: 1-50 m. Ge: A coral reef species more often found on protected than on exposed reefs, fairly secretive. Below: **Black-saddle grouper** *Plectropomus laevis,* to 100 cm, entire area, 10-150 m, variable, juveniles mimic the poisonous pufferfish *Canthigaster valentini.*

Anyperodon leucogrammicus **Banda Sea, Indonesia**

Spotted grouper
Le: To 60 cm. Di: Entire area. De: 3-40 m. Ge: A coastal species, preferring algae reef habitats in protected bays and lagoons. Spots in juveniles proportionally larger, becoming elongate on head of large adults. Below: **Vermiculated grouper** *P. oligacanthus,* to 70 cm, entire area, 3-50+ m.

Plectropomus maculatus **Derawan I., Indonesia**

Yellowfin grouper
Length: To 75 cm. Distribution: Andaman Sea to Bali. Depth: 10-150 m. General: Juveniles (small photo below) in shallow water, adults on deep reefs. The blue coloration of the body can be changed from deep to light blue. Larger adults are entirely spotted white.

Epinephelus flavocaeruleus **Singapore**

Epinephelus cyanopodus **Ogasawara Is., Japan**

Purple grouper
Le: To 75 cm.
Di: West Pacific, throughout our area. De: 2-150+ m.
Ge: Highly variable, small juvenile (below) dark blue with orange tail, changing to light grey-blue with numerous spots in large adults. Fins may be yellow. Coastal reefs with caves and large silty lagoons.

Epinephelus ongus **Sulawesi**

White-speckled grouper
Le: To 35 cm. Di: Entire area. De: 3-30 m. Ge: Clear coastal bays and on mixed coral and rock substrates. Secretive along drop-offs but often at entrance of ledge or cave. Sometimes on large sponges. Below: **Marbled grouper** *E. macularis,* to 65 cm, entire area, 2-100 m, in lagoons.

Epinephelus hexagonatus **Okinawa, Japan**

Hexagon grouper
Le: To 30 cm. Di: Entire area. De: 4-48 m. Ge: Pattern distinct. Below: **Spot-snout grouper** *E. polyphekadion,* to 65 cm, entire area, 3-50 m, coastal bays and protected inner reefs from shallow reef flats to deep lagoons and along the base of drop-offs. Best recognised by the two spots on the snout.

Magenta slender fairy basslet

Length: To 7 cm.
Distribution: Entire area.
Depth: 1-30+ m.
General: This fairy basslet is often abundant where it occurs. The tiny fish swarms around sea fans and soft corals. This genus does not show obvious differences between males and females, though we notice colour patterns, evident in the photo, with and without the red flash near the tail.

Fairy basslets often occur in large aggregations which live more or less stationary in a certain area of the reef, usually an exposed section that provides sufficient food in the form of plankton drifting past. Almost all of the fairy basslets are very colourful and thus add a special note to Asian coral reefs. Understandingly, they are diver's delight and provide more than one opportunity to watch and photograph them against the background of a healthy reef.

Luzonichthys waitei **Talaud I., Indonesia**

Flame fairy basslet

Length: To 9 cm.
Distribution: Malaysia, Singapore, Indonesia, Philippines.
Depth: 1-15 m.
General: This species is found on clear coastal to outer reefs in current-prone areas. It usually occurs in large schools along the upper edge of drop-offs. Males are often busily displaying to each other or to the opposite sex by showing their red dorsal fin.

The photo at the right also shows some Jewel fairy basslets *P. squamipinnis*. This species is treated on the following page.

Pseudanthias dispar **Nusa Penida, Indonesia**

Pseudanthias squamipinnis **Flores, Indonesia**

Jewel fairy basslet

Length: To 15 cm.
Distribution: Widespread throughout the Indo-Pacific, including all of our area but has some geographical variations.
Depth: 3-20 m.
General: On reef crests, slopes and drop-offs, often in large aggregations, especially where the dominant species of the genus. Most reef habitats from silty coastal to clear outer reefs, along drop-offs and in lagoons. Colour highly variable from red, orange to purple and in some areas peppered with yellow spots. Females orange.

Pseudanthias huchtii **Palawan, Philippines**

Pacific fairy basslet
Le: To 9 cm. Di: Indonesia to Philippines. De: 3-20 m. Ge: Fairly common, in small groups on reef crests, along upper region of drop-offs. Rich reefs with mixed stony and soft corals. Male with red stripe on cheek. Below: **Randall's basslet** *P. randalli,* to 7 cm, entire area ,15-70 m. Mostly deep.

Pseudanthias bicolor **Yap**

Yellow-back basslet
Le: To 13 cm. Di: Entire area. De: 20-70 m. Ge: In small groups along deep outer reef drop-offs but also in clear coastal areas. Usually in 30+ m where back looks bright yellow in natural light. Below: **Two-spot basslet** *P. bimaculatus,* to 14 cm, Bali and Flores, Indonesia, 10-70 m. Deep reef slopes.

Pink fairy basslet

Le: To 10 cm. Di: Entire area. De: 6-50 m. Ge: Pale to darker pink, juveniles with red tips and narrow margin on caudal fin. During display male has pale forehead and red caudal fin. Below: Similar **Luzon fairy basslet** *P. luzonensis,* to 12 cm, Philippines, Indonesia, New Guinea, 20-60 m.

Pseudanthias hypselosoma **Mabul I., Malaysia**

Mirror fairy basslet

Le: To 15 cm. Di: Entire area. De: 10-150+ m. Ge: Coastal to outer reef drop-offs. Usually very deep, over 30 m, but in some places at snorkelling depth. Often in large groups of mixed sexes feeding well out in the open along drop-offs. Highly variable, blotch of male (right) may be absent. Below: Female.

Pseudanthias pleurotaenia **Bunaken I., Indonesia**

Red-band fairy basslet

Le: To 10 cm. Di: Entire area. De: 10-60 m. Ge: Variable pale to dark pink, juv. with red tips and narrow margin on caudal, males develop broad red band over mid-body. On isolated rocks along deep slopes. Below: Similar **Cooper's fairy basslet** *P. cooperi,* to 14 cm, entire area, 5-60 m.

Pseudanthias rubrizonatus **Batangas, Philippines**

Lori's basslet

Le: To 8 cm. Di: West Pacific, mainly Indonesia, ranging to Philippines and Australia. De: 15-70 m. Ge: Found mainly along clear water drop-offs in caves with black corals and large fans, but also in outer reef lagoons in rich coral areas. Left: Female group. Below: Male.

Pseudanthias lori **Sulawesi**

Purple queen

Le: To 11 cm. Di: West Pacific, Indonesia to Philippines. De: 10-25 m. Ge: Clear coastal to outer reefs along drop-offs and slopes in current-prone areas. May form large aggregations to feed well above the substrate. Females have a yellow stripe along base of dorsal fin. Left: Female group. Below: Male.

Pseudanthias tuka **Nabucco I., Indonesia**

Sailfin queen

Le: To 15 cm. Di: Entire area. De: 5-60 m. Ge: Rare in Indonesia and only seen north of Sulawesi, mainly from S-Japan ranging along E-New Guinea to Australia and further east into Pacific. Forms large schools on outer reef slopes, feeding high above the substrate. Left: Female group. Below: Male.

Pseudanthias pascalus **Kerama I., Japan**

Snowflake soapfish

Le: To 34 cm. Di: Entire area. De: 15-150+ m. Ge: Secretive, but common in certain areas. Prefers small islands where in loose groups under corals. Beard-like skin flap, spot on tip of snout distinct. Below: **Yellow emperor** *Diploprion bifasciatum*, to 25 cm, entire area. De: 1-50 m.

Pogonoperca punctata **Cebu, Philippines**

Lined soapfish

Le: To 25 cm. Di: Entire area. De: 1-150+ m. Ge: In a variety of coastal habitats from shallow estuaries, lagoons, rock pools to deep muddy slopes. Young usually shallow. Large adults in which the lines brake up into dashes appear to be restricted to deep water.

Grammistes sexlineatus **Banda Sea, Indonesia**

Arrow-head soapfish

Length: To 15 cm.
Distribution: Indo-Pacific, all of our area.
Depth: 10-50 m.
General: A solitary species commonly found found in caves along drop-offs of coastal as well as outer reefs. Due to its preference for dark places this soapfish is often missed by divers, except when they use underwater torches or lights. The species is easily identified by its general shape and coloration.

Belonoperca chabanaudi **Irian Jaya**

A LAKE IN THE SEA

The recommendations of scientists couldn't be overheard by the author: "If you dive the east coast of Kalimantan/Borneo, don't forget to visit the island of Kakaban. It is the attraction in the area because it is absolutely unique. You will find a biological paradise for in its centre lies a fascinating lake without any connection to the sea that surrounds Kakaban. It is a lake full of jellyfishes, but don't be afraid, they don't sting - at least not in each case. Compared to Kakaban the well-known jellyfish lake of the Palau Islands is a biological desert!"

PHOTOS: GERALD ALLEN

An aerial view of the island of Kakaban.

Kakaban itself is uninhabited, but is a regular diving spot accessible from the neighbouring diving centres in the islands of Derawan, Sangalakki, and Nabucco because on a steep outer reef drop-off of Kakaban lies Barracuda Point where over 100 of these predatory fishes schooling together are constantly present. Down there in a depth of 25 metres a strong current prevails most of the time, but to compensate for the drag one can also see formations of horse mackerels, snappers, and often also batfishes swimming alongside the barracudas in the current. Knowing that large fishes have become rare in Indonesian waters these days such a diving location is especially appreciated!

We had the barracudas on film who had lastingly fascinated our diving party. But now I urged the diving guide to proceed to our true destination for this day. The calmness of the bay where we anchored was a little bit betraying because reaching the jellyfish lake became quite troublesome. The steep karst cliffs of the approximately 50 m high rim around Kakaban had to be climbed and the path across fallen and rotting trees was very slippery too. Clutching at our cameras it took us about half an hour until we had reached the muddy shore of the lake. Only rarely I had felt the weight of the diving gear in the damp climate of the equatorial jungle so much as on that occasion.

Underwater visibility in the lake is considerably less when compared to that of the surrounding sea, but with 8-10 m still acceptable. In comparison to the coral reef habitat around Kakaban species diversity is very poor, however, this penury is compensated for more than enough by the unique ecology of the lake and the unusual mutual relationships between many of its inhabitants. This lake offers the rare opportunity to observe plants and animals which normally occur only in highly polluted river mouths where diving is impossible.

The prevailing animal species are jellyfish, especially two species thereof. In the deeper areas a species of rhizostome jelly forms dense concentrations. It's like swimming in a kettle of jel-

Right: At last the divers have reached the shore of Lake Kakaban. Facing page: The margins of the lake thrive with all kinds of life, mangroves on the land, sponges and algae under water.

ROGER STEENE

HELMUT DEBELIUS

Cassiopeia jellyfish on a carpet of *Halimeda* algae.

lyfish soup. It's impossible to swim without bumping into them. Although they usually do not sting, I wear my wetsuit to prevent stings. Another type of jellyfish belonging to the genus *Cassiopeia* lives on the bottom right up to the surface. The members of this group are resting on the bottom - sand or algae - upside-down with their tentacles directed upwards. They cover some sections of the lake like a solid carpet! What do these myriads of cnidarians feed on? Their greenish-brown colour indicates the presence of zooxanthellae. These unicellular algae also live within the cells of corals and utilise sunlight and dissolved nutrients to produce energy-rich sugars that are "leaked" to their hosts. One of the

In Lake Kakaban the goby species *Exyrias puntang* feeds in groups on single jellyfishes.

most unusual feeding associations in the lake involves a yet unidentified species of stalked anemone. In the absence of their usual food source - plankton - the Kakaban anemone has evolved into a jellyfish predator. This must be seen to be believed - the jellyfish victim is frequently larger than the anemone.

In spite of Lake Kakaban's relatively large size of 5 km^2, only 8 fish species were found by scientists. The surface waters are populated by halfbeaks (Hemiramphidae). The common name of these slender fishes alludes to the peculiar mouth structure in which the upper jaw is only a fraction of the size of the very elongate lower jaw. Together with schools of silversides (Atherinidae) they search the surface for insect food. The cardinalfish *Apogon lateralis* and the goby *Exyrias puntang* are by far the most abundant fishes. Both are less than 15 cm long, and well-known to inhabit the mangrove shorelines of the Indo-Malay Archipelago. Usually they are shy and occur in far fewer numbers than in the lake. Due to the absence of predators, their timid behaviour has altered: Totally fearless they let themselves be photographed, and even follow me. I am very surprised when on one occasion I witness a pack of gobies attacking a jellyfish.

The cardinalfish *Apogon lateralis* is one of the rare species in the lake.

Harmless file snakes (*Acrochordus* spp.) hunt under water for some minutes only.

The ecology of the lake is a study of basic simplicity. Plants and animals that are numerous in most marine systems are poorly represented with the exception of sponges. The latter grow in colourful colonies attached to the mangrove roots at the edge of the lake. It is an unusual diving experience to sink into the algae carpets on the bottom at a depth of 6 m or on the algal reefs near the mangrove forest. There I encounter a sea snake, but later I am informed that it is a non-poisonous member of the file snake family (Acrochordidae) which leaves dry land to hunt the small fishes in the lake. The lake's shoreline is easily its most attractive feature. It is fringed with a verdant band of mangrove trees, their tangled prop-roots forming an impenetrable maze at the lake's edge. It is here that the photographer finds colourful tunicates, sea cucumbers, diverse algae, and the sponges already mentioned before. While diving down I note that sand is replaced by mud bottom on which many brown brittle stars live. Although the lake is not deep, after some time the air in my diving tank is finally used up. The films in both of my cameras are exposed long since anyway.

Scientists presume that a certain amount of water exchange exists across the lake's bottom which, however, is certainly not sufficient to maintain the wealth of marine life at a salinity of 24-26 parts per thousand, considerably lower than the surrounding seas. Most probably the important nutrients have their origin in the lake and are effectively recycled. Algae and phytoplankton inhabit the sun-lit areas and are responsible for the high oxygen level. They are the so-called primary producers and the basis of the food chain. *Halimeda* and *Caulerpa* algae build up a "reef maze" in the lake; they are able to take up calcium carbonate and deposit it in their leaf-like organs.

The origins of Kakaban Island are still obscure, but the idea is that is was an atoll that has been geologically uplifted. As a result the former coral lagoon is now a land-locked lake which occupies most of the interior of the 5-km-long uninhabited island. It is 2.5 km long, and 1.4 km wide. The lake contains a mixture of sea and rain water, has a maximum depth of 11, and an average depth of 4 metres.

A rare document: Smallish sea anemones prey on considerably larger jellyfish.

A large family of small and often colourful fishes. Presently at least 6 genera and over 70 species are recognised, but this complicated family of small and secretive fishes is under revision with many new species recently discovered in the south-east Asian region. Most species are placed in *Pseudochromis* but this genus comprises several distinct groups and is only provisionally used for some of the species included here, as they are waiting to be assigned to new genera in a future family revision.

Dottybacks are commonly found on all kinds of reefs with numerous small hiding places such as cracks in rock faces, boulders, small caves or old coral skeletons overgrown by new ones, also in mixed algae, sponge and coral habitats. Some species prefer silty coastal reefs and others exclusively live along pristine outer reef walls. Most species are seen singly but their companion is probably only out of sight at the time of the observation. Some species live in small aggregations of mixed sizes and may have a large cave as a territory, occupying the narrow ledges and numerous narrow passages which are typical in coral reef drop-offs. They are small hunters, looking for almost anything which moves, including most crustaceans and small fishes. Some of the larger species can be very aggressive and are not suited for a community aquarium.

Some species are known to produce clusters of eggs which are guarded by the male. Hatchlings are about 2.5 mm long and postlarvae settle when about 15 mm long. Juveniles are less colourful than adults and in some species the sexes have different colours. The colour variations within one species can be unbelievably wide ranging, changing with habitat and also geographically, to such extend that one would regard them as different species.

Pseudochromis splendens **Alor I., Indonesia**

Splendid dottyback

Length: To 13 cm.
Distribution: Philippines to eastern Indonesia.
Depth: 3-35 m.
General: A common species in Flores, Indonesia, where these fishes occur on coastal inner reef slopes with rich coral and sponge growth. Tiny juveniles are mostly yellow and often seen in small groups in stringy to bushy sponges which provide good shelter. Adults often live in large yellow tubular sponges.

Purpleback dottyback

Le: To 6 cm.
Di: Malaysia, Philippines.
De: 5-25 m.
Ge: Lives in small groups among corals and rocks on reef slopes and bases of small drop-offs. A very shy species, but the most attractive dottyback in the region.

Pseudochromis diadema **Batangas, Philippines**

Two-tone dottyback

Length: To 5 cm. Distribution: Indonesia and New Guinea. Depth: 10-50 m.
General: Easily identified. In some areas commonly found along sponge- and coral-rich drop-offs along ledges and in small caves, ranging from coastal to outer reef habitats.

Pseudochromis paccagnellae **Ambon, Indonesia**

Magenta dottyback

Le: To 6 cm. Di: Philippines to Samoa and S-Japan. De: 6-65 m. Ge: Uniformly but brightly coloured. Swims out in the open but close to the substrate. Appears blue when seen in natural light. Occurs on outer reef walls and often on the sides of channels near ledges and small caves.

Pseudochromis porphyreus **Palau**

Pseudochromis bitaeniatus Flores, Indonesia

Slender dottyback

Length: To 12 cm.
Distribution: Indonesia to Philippines.
Depth: 3-30 m.
General: Coastal to outer reef slopes and drop-offs with dense coral and sponge growth, moving through passages and staying very close to the substrate. There is little variation in the coloration of various sizes or both sexes, the white band easily identifies this species.

Pseudochromis polynemus Palau

Long-finned dottyback

Length: To 12 cm.
Distribution: Indonesia from Sulawesi north, Philippines, Palau.
Depth: 10-30 m.
General: Along clear coastal and outer reefs with lots of sponges, especially long grey tubular species in which they often disappear out of sight. Swims in an unusually way by strongly twisting its tail with short dashes through the passages which makes it almost impossible to get a good side-on picture.

Pseudochromis steenei Bali, Indonesia

Lyretail dottyback

Length: To 12 cm.
Distribution: Indonesia.
Depth: 8-50 m.
General: This species was recently described as distinct from the similar *P. moorei* on the basis of its coloration, from a few specimens from Bali. However in Bali alone it is highly variable with males partly to completely orange, sometimes yellow tail, and females dark with yellow or dark tail similar to male. Around Bali on deep sand slopes with rubble ridges and small coral or rocky outcrops with crinoids or soft corals. Left: Male at top.

Dusky dottyback

Length: To 10 cm.
Distribution: Indo-Pacific, throughout our area.
Depth: 1-30 m.
General: A common and one of the most variable species, ranging from drab grey or brown to bright yellow. Sometimes yellow over the back. Found in almost every reef habitat from shallow coastal lagoons and seagrass beds with reef patches to outer reef lagoons.

The fish shown on the right was observed for almost 15 minutes hovering above an empty bivalve shell and it seemed like the fish was blowing water into it. The author had never seen an egg deposit of a pseudochromid and was amazed when turning the shell and looking at one (centre photo). Of course, he returned the shell to the same position in which he had found it. The guarding parent fish soon was hovering above it again.

Pseudochromis fuscus **Batangas, Philippines**

Pseudochromis fuscus **Batangas, Philippines**

Firetail devil

Le: To 20 cm. Di: Malaysia, Indonesia, Philippines and southern Japan. De: 1-20 m. Ge: Variable from greenish grey to bright red, more or less barred on the sides. In channels on reef crests among overgrown coral rocks. Male with red tail.

Labracinus cyclophthalmus **Ogasawara Is., Japan**

LONGFINS — PLESIOPIDAE

Calloplesiops altivelis **Flores, Indonesia**

Calloplesiops altivelis **Palawan, Philippines**

Comet

Le: To 20 cm. Di: Widespread in tropical Indo-Pacific, entire area. De: 3-50 m. Ge: Number of spots increases with age as they become proportionally smaller. In large adults the fin spots become elongate, eventually join and form lines beginning at the outer margin. Although common it is not often seen unless one uses a torch when diving along drop-offs. Found among old coastal collapsed reefs in muddy areas as well as on outer reefs deep in the back of caves and under ledges. Usually only one is seen but as many as three may live together in a small part of a reef less than one square metre (Flores). In Japan several pairs were observed in a shallow reef with long ledges.

Longfins are a small family comprising 6 genera and about 20 species. Except for some large and spectacular temperate species, they are small and very secretive. The Comet is a popular aquarium fish which breeds in captivity. It is one of the larger tropical species which is very widespread in the tropical Indo-Pacific. Other species are mouth brooders in which the male incubates the eggs. There are several other species of the genus in our area but they are small, very secretive and only rarely seen. They are mostly drab with some blue markings and not included here.

BIGEYES — PRIACANTHIDAE

Priacanthus hamrur **Komodo I., Indonesia**

Common bigeye

Le: To 40 cm. Di: Entire area. De: 15-250 m. Ge: Common on outer reef slopes. Often up to 100 under ledges of coral reefs. Feeds at night on plankton. Below: **Blotched bigeye** *Heteropriacanthus cruentatus*, to 35 cm, entire area, 3-30 m.

Pterapogon kauderni **Sulawesi**

Pterapogon kauderni **Sulawesi**

Banggai cardinalfish

Le: To 7 cm. Di: Only known from the Banggai Islands, central-east Sulawesi, Indonesia. De: 0-2 m. Ge: Probably the most distinctive family member with a unique coloration. This species does not incubate its eggs in the mouth like other cardinalfishes but harbours its young in the mouth for several days after they have hatched. For apogonids the eggs are large, 2.5 mm in diameter, and few in number, only about 20. The limited distribution of the species obviously has to do with its extreme specialisation concerning its reproduction. Schools near long-spined sea urchins in shallow seagreass habitats. Juveniles seek protection near anemones. See also previous page.

The cardinalfishes are a very large family of small reef fishes, most numerous in tropical waters. They have a global distribution, 26 genera contain an estimated 250 species. The majority lives in shallow coastal waters in caves and crevices during the day, drifting out during the night to feed. A few live deep on continental slopes, some are restricted to freshwater. Their diets consists of small plankton but a few feed on small fishes. Apogonids have two spines in their anal fin and two separate dorsal fins. The reef dwellers are marked with stripes and spots.

Apogon bandanensis **Sangihe I., Indonesia**

Banda cardinalfish

Le: To 10 cm. Di: Entire area. De: 3-30 m. Ge: Yellow margins on caudal fin and on soft dorsal and anal fin tips, body banding more defined in adults. Coastal lagoons and reefs. Below: **White-spot cardinalfish** *A. dispar,* 6.5 cm, entire area, 10-60 m. Distinct white spot above a black spot on caudal fin base.

Ring-tail cardinalfish

Le: To 14 cm. Di: Entire area. De: 1-50 m. Ge: Adults distinctly marked with black band on caudal peduncle which in small juveniles appears as a large spot. Coastal reef crests and slopes, usually in small to large aggregations near large overhangs. Adults pair within aggregations.

Apogonids are one of the few marine fish families which practise mouth brooding. After fertilisation the male takes the eggs in its mouth for incubation which lasts just over one week in tropical waters and slightly longer in cooler zones. Reports of females incubating eggs appears to be based on males actually eating a brood before taking on a second brood, thus appeared to be gravid during incubation. The hatching larvae have a long planktonic stage (about 60 days) dispersing them over great distances. Apart from the brooding activity sexes are very similar. In some species the female is larger and more colourful, particularly during courtship which she dominates. Most species pair when adult, including many schooling ones.

Apogon aureus **Sipadan I., Malaysia**

Apogon aureus **Singapore**

Goldbelly cardinalfish

Le: To 10 cm. Di: Entire area. De: 10 - 35 m. Ge: Singly or in pairs. Lives hidden in caves and crevices in the shaded reef. Below: **False three-spot cardinalfish** *A. rhodopterus*, to 14 cm, Indonesia and New Guinea, probably wide-ranging, 3-30 m. Secretive on clear coastal reefs, active at dusk.

Apogon apogonoides **Tioman I., Malaysia**

Apogon sealei **Okinawa, Japan**

Cheek-bar cardinalfish

Le: To 10 cm. Di: Indonesia and Philippines. De: 3-25 m. Ge: Similar to *A. chrysopomus* but spots replaced by bars on opercle, absent on cheeks. Found in same region but schools among branching corals, seems to prefer clearer waters. Below: **Cheek-spots cardinalfish** *A. chrysopomus*, to 11 cm.

Apogon leptacanthus **Gulf of Thailand**

Longspine cardinalfish

Le: To 6 cm. Di: Entire area. De: 2-15 m. Ge: In groups in the surf zone of reefs and lagoons. Below: **Rifle cardinalfish** *A. kiensis,* to 8 cm, widespread tropical Indo-Pacific, entire area, ranging into warm-temperate zones, 6-50 m. Distincly striped. Dusky area on anal fin base.

Apogon angustatus **Bunaken I., Indonesia**

Blackstriped cardinalfish

Le: To 8 cm. Di: Entire area. De: 10-25 m. Ge: Singly or in pairs. It occurs hidden in caves and below overhangs. Below: Similar **Black-striped cardinalfish** *A. nigrofasciatus*, 10 cm, widespread tropical Indo-Pacific, including all of our area, 3-50 m. Coastal to outer reef drop-offs and rocky estuaries.

Orange-lined cardinalfish
Le: To 9 cm. Di: Indo-Pacific, including all of our area. De: 1-35 m. Ge: Several similar species confused under this name. Below: **Multi-striped cardinalfish** *A. multilineatus*, to 10 cm, widespread tropical west Pacific, entire area, 3-25 m. Numerous thin black lines, face black with thin pale lines.

Apogon cyanosoma **Bali, Indonesia**

Five-line cardinalfish
Le: To 10 cm. Di: Entire area. De: 3-40 m. Ge: Distinctly marked, juveniles more yellow on caudal peduncle. Coastal reef crests to outer reefs. Below: **Tiger cardinalfish** *C. macrodon,* to 22 cm, entire area, 3-30 m. Coastal reefs and rocky estuaries. A bit of yellow on head remains in adults.

Cheilodipterus quinquelineatus **Flores, Indonesia**

Pyjama cardinalfish
Le: To 8.5 cm. Di: Entire area. De: 3-25 m. Ge: Inshore, in lagoons, in small to very large groups amongst dense branching corals. Below: **Polkadot cardinalfish** *S. orbicularis*, to 10 cm, widespread tropical Indo-Pacific, including all of our area, lives close to the surface. Inshore, in mangroves, lagoons.

Sphaeramia nematoptera **Nabucco I., Indonesia**

TINY FISHES – A PSYCHIATRIST'S VIEW

Mitsuaki Takata was born in Kobe, Japan, in1966. He began photographing in 1983, and started SCUBA diving and underwater photography in 1993. In the following year, he moved to Okinawa Island at the southwest end of Japan where Japan's most representative diving area is situated. He began super-macro photography in 1995. As a photo artist (and psychiatrist) he developed his own, very special philosophy concerning his work.

PHOTOS: MITSUAKI TAKATA

Portrait of a Red-spotted blenny *(Blenniella chrysospilos)*. Okinawa I., Japan, depth 1 m, length about 12 cm.

"Communicability is the most important factor in my underwater photography. Plants have marks on their petals that reflect ultraviolet light and thus indicate the location of the nectar to insects which pollinate them. When danger is approaching, partner gobies relay that information to the alpheid shrimps that share their burrow with the fishes. Cleaner wrasses and cleaner shrimps dance in peculiar ways to advertise to other fishes that they are cleaners. Men communicate actively just like other mammals such as dogs, cats and dolphins. Some people even try to communicate with snakes or chameleons. My super-macro photography from a tripod is the communication device between very small fishes and you. You can read the mind of tiny underwater living things with the help of my photos.

I am very interested in facial expressions of nature's creatures. Having been affected by my grandfather, who was a sculptor, I am fond of painting since my childhood. My favourite topic was the portrait. I began photographing at the age of 17, and again my favourite were portraits. The interest in facial expressions plays an important role in my profession as a psychiatrist. And this is also reflected in my underwater photography. Gradually, I became more and more interested in tiny creatures and at some point found out that they also have a rich repertoire of facial expressions, but that their faces were too difficult for us to read. I wanted to manifest these expressions somehow. That was the clue to my further work.

I take photographs to 'talk' with tiny fishes through the viewfinder of my super-macro camera. In such a way I am an underwater portraitist of small creatures. It would be a great pleasure for me if those tiny creatures appeal to you through my photos".

Blue-striped cave goby *(Trimma tevegae)*. Okinawa I., Japan, depth 15 m. This is a small goby which is common in Okinawa. It prefers a dark environment. I used special lighting when taking this photo to express the vague fear inside a dark cave. One of the strobes gave red light. The actual body colour is a graceful gold.

Right: Portrait of a Yellownose partner goby *(Stonogobiops xanthorhinica)*, Okinawa I., Japan, depth 20 m, length about 5 cm. These little fish live together with burrowing alpheid shrimps. Both animals have a mutual relationship, the shrimp (usually a pair) building the home, and the fish (sometimes also a pair) guarding it. People do the same when keeping dogs as guarding companions.

Below: Profile of a Purple fire goby *(Nemateleotris decora)*. Sipadan I., Malaysia, depth 22 m, length about 7 cm. Violet is my favourite colour. Hence I consider this species one of the most beautiful gobies, but of course it shows little colour at depth. It's a very shy species and in Japan usually lives at depths greater than 40 m. This is too deep to take super-macro photos! Fortunately, I could find an individual in only 20 m near Sipadan Island. Every time I visit there, I always dive to see this individual and take photos. It's very difficult to photograph it for it always hovers above its burrow.

Left: Yano's or Flagtail partner goby *(Amblyeleotris yanoi)*. Okinawa I., Japan, depth 35 m, length about 9 cm.

This goby has the most refined colours among all species of the genus *Amblyeleotris*. Pearl white and pale gold - this simple colour combination makes me feel like a young nobleman of the depth. The tender-looking eyes appear gentle to us humans.

Below: A Lemon coral goby *(Gobiodon citrinus)* on a Pillow sea star. Okinawa I., Japan, depth 7 m, length about 3 cm.

The Pillow sea star is very colourful in the Okinawa region and often has tiny shrimps on it. But we don't see gobies on them. When I dived at night, I stumbled upon a Coral goby on a Pillow sea star. This goby usually lives among branches of *Acropora* corals, and seldom leaves them. But during the night, I sometimes see it out in the open. At that time it was not shy. Perhaps it was sleepy. So I was able to take this shot.

Right: This tiny Lemon coral goby *(Gobiodon citrinus)* lives among the branches of *Acropora* corals just like a bird which lives on a tree. Okinawa I., Japan, depth 5 m. The size of this fish is very small, only 6 mm. To see its facial expression, a super-macro photo system is the best way. It is seen here resting on the branchial spire of a tubeworm residing in the coral.

Below: The first dorsal fin of Randall's partner goby *(Amblyeleotris randalli)*. Mabul I., Malaysia, depth 20 m, length 8 cm.

For me, the first dorsal fin of this goby is one of the most beautiful fins of all fishes. It shows a subdued black spot and a white and yellow flare around it. The whole arrangement looks like a solar eclipse, "sensu" in Japanese. When I photographed this fin I concentrated on its marking, so I did not regard the actual direction of the fish's body as important. The direction of the goby's head is to the lower left of this photo. It's very difficult to take such a photo because this goby doesn't spread its first dorsal fin on command. I tried to approach carefully and waited for my chance. Pattern and colours remind me of Japanese traditional colours.

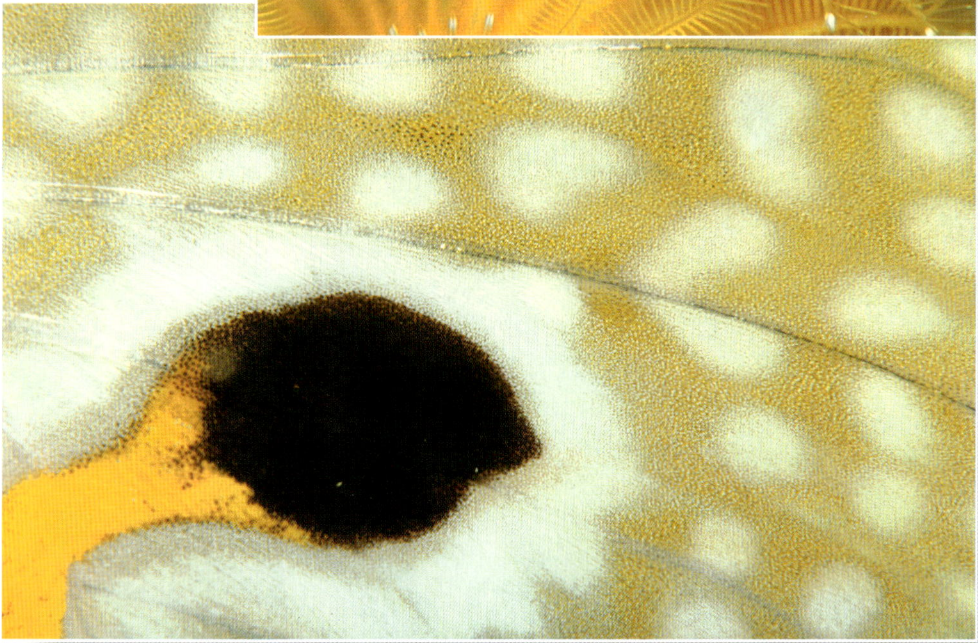

A small tropical family with representatives worldwide, comprising two genera and nine species. The genus *Hoplolatilus* with the majority of species is confined to tropical coral reefs. The two Indo-Pacific species of *Malacanthus* are widespread and range into the warm temperate zones. They are commonly known as blanquillos. These slender fishes are found on sandy reef flats, either singly or in pairs, swimming typically just above the bottom and dashing over short distances for a quick stop to study their surroundings. Burrows are made under rocks on sand and often several are in the vicinity of the fishes' feeding range where they can quickly seek cover from a potential threat. Adults may build large nesting sites, shifting large quantities of sand. They are distinctly marked with stripes on the body or tail, juveniles of some species differ greatly from adults. Eggs are probably laid in their large rubble mounds which sometimes measure more than 1 m in diameter and are guarded by a pair. Hatchlings are pelagic, and postlarval stages are large when settling, 3-7 cm, suggesting a prolonged pelagic life.

Malacanthus latovittatus **Bali, Indonesia**

Blue blanquillo
Le: To 40 cm. Di: Entire area. De: 6-40 m. Ge: Juveniles very different from adults. Adults usually in pairs on outer reef crests, juveniles on coastal reef slopes, often quite deep. Below: **Flagtail blanquillo** *M. brevirostris*, to 30 cm, entire area, 6-50+ m. Caudal fin rounded in juveniles, more truncate in adults.

Hoplolatilus fourmanoiri **Nha Trang, Viet Nam**

Yellow-spotted tilefish
Le: To 14 cm. Di: Philippines, Vietnam. De: 5-70 m. Ge: Described from off Vietnam. Prefers silty sand bottoms with coral rocks. Below: **Yellow tilefish** *H. luteus*, to 10 cm, Lesser Sunda Islands, Indonesia, 15-35 m. Only known from Indonesia. On muddy bottoms in channels.

Red-stripe tilefish

Le: To 12 cm. Di: N-Indonesia to Japan. De: 30-80 m. Ge: In pairs, deep on sand and coral rubble. Shy and difficult to take photos of. Builts large mounds as nesting sites. Below: **Chameleon tilefish** *H. chlupatyi*, to 11 cm, Philippines, 20-40 m. In pairs. Can change colour from light green to dark blue.

Hoplolatilus marcosi **Yonaguni I., Japan**

Purple tilefish

Length: To 15 cm.
Distribution: Papua New Guinea to Philippines.
Depth: 20-70 m.
Ge: A secretive species living on sandy and coral rubble bottoms and mainly found below a depth of 40 m. Usually it is seen in pairs. One partner of these plankton feeders swims some metres above their mound while the other moves around at the entrance to their burrow.

Hoplolatilus purpureus **Cebu, Philippines**

Blue-head tilefish

Le: To 15 cm. Di: West Pacific, entire area. De: 20-50+ m. Ge: Found along the slopes on the bases of reef drop-offs. In most areas rarely seen in less than 30 m. Inhabits coastal as well as outer reefs, usually in pairs and staying close to burrows. Below: Juvenile.

Hoplolatilus starcki **Tukang Besi, Indonesia**

REMORAS — ECHENEIDIDAE

Echeneis naucrates — **Talaud I., Indonesia**

Slender suckerfish
Le: To 100 cm. Di: All tropical seas, except eastern Pacific. De: Surface to very deep, depending on host. Ge: Mostly pelagic offshore, but enters coastal and estuarine waters with host. Usually attached to sharks and rays, often swimming free.

Remoras are a small family (all are easily recognised by the sucking disc dorsally on the head and nape), comprising 4 genera and 8 species. The sucker disc is the modified first dorsal fin, consists of transverse movable laminae and is used to attach by suction to other fish, turtles and whales.

Remorina albescens — **Gulf of Thailand**

White remora
Length: To 30 cm. Distribution: Entire area. Depth: 2 - 30 m. General: Occurs more in the open water column, prefers Manta rays as hosts (attached near mouth or gills).

Some remoras are quite host-specific, others, however, may attach to about anything which moves including ships and divers. Food is obtained when the host is feeding, but parasitic copepods attached to the host are also taken. Remoras feature elongate bodys and a depressed head with lower jaw greatly protruding. Dorsal and anal fins are similar in size and lie opposite each other.

COBIAS — RACHYCENTRIDAE

Rachycentron canadum — **Bali, Indonesia**

Cobia
Le: To 200 cm. Di: All tropical seas, except east Pacific. De: 2-50 m. Ge: Juveniles distinctly striped, becoming plain brown in large adults. Juveniles usually singly, adults ofte sighted in small groups of 3-6 individuals, also entering shallow water.

The family contains a single wide-ranging species which is pelagic but occasionally visits reefs and under water has the appearance of a shark which can be scary for divers because these fishes are curious and often approach very quickly and within close range.

A small tropical family, comprising 4 genera and 20 species, some ranging into warm temperate zones, but restricted to the Indo-Pacific region. They are closely related to the snappers (family Lutjanidae, p.112) but are schooling, planktivorous, smallish fishes with streamlined bodies. On the reefs these fishes have a high ornamental value with their bright colours and orchestrated movements, as large schools commonly patrol the reef slopes and walls in pursuit of zooplankton. Differences between genera are mainly found in the anatomy of the mouth region and the three genera included in this family are difficult to distinguish from external features alone. Usually *Caesio* is deeper-bodied and more robust while *Gymnocasio* is the most slender genus. The genus *Pterocaesio* is variable in shape from slender to moderately deep-bodied.

Egg and larval development is identical to that of the lutjanid snappers with small pelagic eggs, less than 1 mm in diameter and hatchlings about 2 mm long. Postlarvae measure about 30 mm in length when settling on the substrate, forming small groups at the base of shallow reefs.

Pterocaesio randalli **Flores, Indonesia**

Yellow-dash fusilier

Length: To 22 cm.
Distribution: Indonesia, Malaysia and Philippines.
Depth: 6-35 m.
General: The bright yellow dash on the sides readily identifies this species. The dash remains visible at great distance and schools appear as yellow flashes against a dark background. Usually near the top of walls along deep outer reefs, in large schools which are nearly always on the move. See also previous page, a huge school, photographed at Sipadan I., Malaysia.

Pterocaesio digramma **Ogasawara Is., Japan**

Black-tipped fusilier
Le: To 30 cm. Di: Entire area. De: 3-25 m. Ge: Distinct by 2 thin yellow lines and black tips on caudal fin. Below: Similar is **Yellow-band fusilier** *P. chrysozona*, to 30 cm, entire area, 3-25 m. Coloration of juveniles and adults slightly different, yellow band broadens with age. Rich slopes, in small groups.

Blue-dash fusilier
Le: To 25 cm. Di: Entire area. De: 6-25 m. Ge: Distinct by broad blue band along body, no black caudal fin tips. At night lower body half bright red. Usually in large schools along deep drop-offs and slopes. Below: **Gold-band fusilier** *Caesio caerulaurea*, to 35 cm, entire area, 3-25 m.

Pterocaesio tile **Bali, Indonesia**

Blue-and-gold fusilier
Le: To 30 cm. Di: Entire area. De: 10-35 m. Ge: Shy, during the day in small groups in caves. Below: **Yellow-back fusilier** *C. xanthonota*, to 30 cm, Java, Indonesia, 2-50 m. Very similar to *C. teres* but yellow extends posteriorly to dorsal fin, sometimes over head. Coastal to outer reefs.

Caesio teres **Sangihe I., Indonesia**

Robust fusilier
Le: To 40 cm. Di: Entire area. De: 3-40 m. Ge: Bright red at night. Several similar species, the others more slender and lacking indistinct stripes below eye, or completely lacking yellow tail. On coastal reefs and in lagoons, usually in large schools. This is one of the larger family members.

Caesio cuning **Derawan I., Indonesia**

A large family comprising four subfamilies, 17 genera and about 103 species worldwide most of which are found in the West Pacific. The largest subfamily are the Lutjaninae with 6 genera and 72 species, 43 of which inhabit the West Pacific. Some doubt exists about the status of the family Caesionidae (p.110) being merely a subfamily of the Lutjanidae. A total of 68 species occur in our area many of which are restricted to deep water while some occur mainly in estuaries and even in freshwater. Included here are the reef dwellers which are commonly observed by divers and snorkellers.

Snappers occur in most benthic habitats and many species change their lifestyle with age, often changing from a life in estuaries or freshwater as juvenile to a full marine environment as adult. They are pelagic spawners, producing a great number of tiny eggs less than 1 mm in diameter. Postlarvae settle when about 20-40 mm long, depending on the species. All are carnivores, feeding on small fishes and invertebrates, either benthic or planktonic. Behaviour varies greatly among family members, some live singly, others are schooling.

Aprion virescens **Layang I., Malaysia**

Green jobfish
Le: To 100 cm. Di: Entire area. De: 10-100 m. Ge: Voracious hunter, high above substrate along slopes. Darts into reef hunting octopus, cuttlefish and also fishes. Large individuals may be ciguatoxic. Very shy to divers. Below: **Small jobfish** *Aphareus furca,* to 40 cm, entire area, 6-70 m. In open water.

Macolor macularis **Luzon, Philippines**

Midnight snapper
Le: To 60 cm. Di: Widespread in our area. De: 5-50 m. Ge: Juveniles singly, postlarvae associated with crinoids, adults may form loose aggregations. Coastal and inner reefs as well as on outer reef drop-offs. Juveniles (below) have extended ventral fins which gradually are reduced with age.

Black snapper

Length: To 60 cm.
Distribution: Entire area.
Depth: 5-90 m.
Ge: Juveniles singly on coastal reefs and in lagoons. Adults form schools along drop-offs on coastal and inner reefs.
Below: Juvenile.

Macolor niger **Banda Sea, Indonesia**

Humpback snapper

Le: To 50 cm. Di: Entire area. De: 1-30 m. Ge: Juveniles usually in mangrove or seagrass areas in the intertidal zone. Adults in often large schools along reef slopes, showing little activity during the day. Very large individuals sometimes singly in shipwrecks. Coloration varies greatly, adults often red.

Lutjanus gibbus **Talaud I., Indonesia**

Red snapper

Le: To 75 cm. Di: Entire area. De: 3-70 m. Ge: Adults in moderate depths, usually singly and just off drop-offs on slopes, often at shipwrecks. Juveniles on reef crests mimicking damselfish when small, hunting prey during the day. Sometimes called Two-spot snapper but spots not distinct.

Lutjanus bohar **Sumatra**

Lutjanus sebae **Bali, Indonesia**

Red emperor
Le: To 100 cm. Di: Entire area. De: 2-100 m. Ge: Small juveniles seek protection between spines of sea urchins living on mud and sand slopes. Gathering urchins may bring juveniles together which form groups before leaving host into deep water. Below: **Timor snapper** *L. timorensis,* to 50 cm, entire area, 0-130 m.

Lutjanus malabaricus **Kerama I., Japan**

Malabar snapper
Le: To 50+ cm. Di: Tropical West Pacific. De: 10-100 m. Ge: Coastal to offshore mud habitats. Juveniles shallow but mainly red adults deep and normally not seen by divers. Below: **Mangrove jack** *L. argentimaculatus,* to 180 cm, entire area, 1-100 m. Small juveniles in mangroves and freshwater.

Lutjanus fulvus **Flores, Indonesia**

Blacktail snapper
Le: To 40 cm. Di: Entire area, including remote islands. De: 1-40 m. Ge: Adults usually singly, deep, in lagoons and on coastal to outer reef slopes. Juveniles (no black tail) intertidal, in freshwater run-offs, may form loose groups. Below: **Lunartail snapper** *L. lunulatus*, to 35 cm, Indonesia to Philippines, 10-50 m.

One-spot snapper

Length: To 60 cm.
Distribution: Widespread in tropical Indo-Pacific.
Depth: 5-30 m.
General: Adults of this species occur in sheltered reefs with caves and are often seen around shipwrecks. Lives mainly singly or in small loose groups which appear to be just sharing a suitable habitat rather than seeking protection in a school. Juveniles are yet unknown and are perhaps simply overlooked because of the similarities to other species or because they occur in deep water - in contrast to those of most other family members.

Lutjanus monostigma **Talaud I., Indonesia**

Blue-striped snapper

Length: To 35 cm.
Distribution: Indo-Pacific, entire area.
Depth: 5-265 m.
General: A versatile species, occurring from shallow to very deep water on coastal as well as outer reefs. Seen in large schools, small aggregations or singly. Juveniles form schools of similar-sized individuals and enter estuaries. The very similar Bengal snapper *L. bengalensis* is mainly an Indian Ocean species but ranges east to Bali, and can be observed swimming together with *L. kasmira*. It is best distinguished by the much paler colour and whitish lower third of the body.

Lutjanus kasmira **Palawan, Philippines**

Lutjanus decussatus **Bali, Indonesia**

Checkered snapper
Le: To 30 cm. Di: Tropical West Pacific. De: 1-30 m. Ge: Juveniles on intertidal reef flats, mimic similar-sized *Halichoeres* wrasses to sneak up on prey. Adults on coastal to outer reef crests. Below: **Half-barred snapper** *L. semicinctus*, to 35 cm, East Indonesia to Philippines, 10-36 m.

Lutjanus biguttatus **Bunaken I., Indonesia**

Two-spot snapper
Le: To 30 cm. Di: Tropical west Pacific. De: 3-30 m. Ge: Coastal reefs and lagoons. In great schools but large adults often singly or in small loose aggregations sheltering under plate corals on outer reef drop-offs. Juveniles often in sheltered lagoons over sand near seagrass, two back spots distinct.

Symphorichthys spilurus **Banda Sea, Indonesia**

Sailfin snapper

Le: To 60 cm. Di: Widespread tropical West Pacific. De: 5-60 m. Ge: Both adults and juveniles are found in similar habitats, sandy flats with coral heads between reefs or in deep lagoons. Large adults develop an almost vertical head profile. Below: Juvenile.

CAMOUFLAGE FOR SURVIVAL

Disguise constitutes one of the most efficient camouflage techniques among some crabs (Brachyura) inhabiting coral reefs and adjacent areas. The Crustacea specialist Arthur Anker tells us more about these fascinating disguise strategies and other types of camouflage in crabs.

PHOTOS: RUDIE KUITER

A crab of the genus *Ethusa* carrying a large sea urchin *Astropyga* sp. Lembeh Strait, Indonesia.

Most common camouflage strategies occurring in decapod crustaceans are: (1) mimicry - imitation of dangerous or dangerously looking animals by colour and usually also shape - for example a non-poisonous crab mimicking a poisonous one; (2) imitation of a part of an animal's environment (dead or living) by the shape of the body - for example a gorgonian-associated crab having the shape of a gorgonian branch extension; (3) imitation of the background colour or colour combination by the colour pattern - for example the same gorgonian-associated crab matching the yellow-orange colour of its host. Indeed, most often cryptic animals combine these latter two tactics.

Decapod crustaceans present probably one of the largest spectrum of camouflage tactics, however, here we deal only with two behavioural patterns typical of Brachyura or true crabs. The first is called "carrying" and the second "decorating". These behaviours are summarised by the more general term allocryptism, defined as active use of various living or dead organisms or their parts and other objects for disguise. While the strategies 1, 2 and 3 are found in many benthic shrimps, anomurans and crabs, allocryptism is limited to Brachyura. Some shrimps are almost transparent, while others combine the transparency with disruptive colour patterns, i. e. patterns which contribute to dissociate the outline of the animal. These latter types of colour patterns are not typical for crabs.

The carrying behaviour has probably a multiple origin since it is found today in two closely related crab families, Dromiidae (sponge crabs) and Homolidae (carrier crabs), but also in Dorippidae (porter crabs) which belong to another section of Brachyura. The first two families represent the most primitive recent crabs. Dromiids and homolids carry for concealment various relatively large objects like sponges (dromiids) or empty shells (homolids, dorippids) with their small fifth legs. These legs are terminated by grasping or holding structures. In dromiids it is a more or less curved, hook-like dactylus. In homolids the two distal articles of the fifth legs are modified to form a small and usually spiny grasping claw (subchela). Dorippid crabs have two last walking legs smaller than the others and adapted for holding of larger objects. The objects carried by crabs are usually large enough to conceal the crab completely. Sometimes, the crabs use unusual, more "exotic" objects for disguise, as documented by these unique underwater shots made by Rudie Kuiter in Indonesian waters. The disguise of these crabs is so extravagant that words like underwater carnival or uw-vogue come to mind!

The camouflage is often so efficient that members of these two crab groups are unidentifiable not only by the predators but also, unfortunately, by the scientists! The photo on this page shows a dorippid crab, probably an *Ethusa* species, carrying a large sea urchin *Astropyga* sp. on the carapace. Interestingly, sea urchins themselves are well known to use vari-

Above & below: This *Dorippe frascone* (Dorippidae) carries a large upside down jellyfish *Cassiopeia andromeda*. Flores, Indonesia.

ous non-living objects (stones, dead algae etc.) for camouflage. The photos on this page show *Dorippe* cf. *frascone* carrying a large upside down jellyfish *(Cassiopeia andromeda)*. It is quite possible that the jelly is still alive. We can guess that the crab came across this jelly which often poses on the bottom and found it a good object for disguise. Indeed, the up-turned umbrella conceals the animal completely, only the eyes remain visible in the frontal view of the animal (bottom photo). It is also possible that the stinging jelly tentacles offer supplementary protection. While most dromiid and homolid crabs carry various objects for disguise, it is not really understandable why some homolid crabs carry cephalopod egg cases! For example, the homolid *Latreillopsis bispinosa* has been reported to carry cuttlefish egg cases in the Philippines. The empty *Sepia* egg cases indicate that the eggs had been carried safely by the crab for a sufficient time to allow the juvenile cuttlefish to develop and hatch.

Decorating behaviour is typical of spider crabs (Majidae) and evolved probably independantly in several related crab subfamilies within Majidae. This family includes some of the most perfect masters of disguise in the Animal Kingom. The carapace and sometimes also the appendages of many majid species are partially covered by setae terminated by a strong hook. Most of these setae are usually inserted just behind the eyes, on the rostrum and near the frontal margin, and each species may have its own insertion pattern. As most objects majids use for disguise have a rough and irregular surface, these hooked setae aid a lot in retaining them on the crab's body. Another evolutionary novelty is a particularly flexible articulation between the articles on the chelipeds of some majid crabs. These articulations enable the claws to be bent backwards and eventually reach the otherwise inaccessible medio-dorsal surface of the carapace. Most majids decorate themselves with pieces of algae, shells and urchins, hydrozoans as well as sea anemones, small pebbles and detritus. The numerous hooked setae of decorator crabs also retain various small organisms (hydrozoans, bryozoans, algae, barnacles etc.) usually called epibionts. Epibionts colonise the crab passively: their juvenile or larval stages settle somewhere on the

The crab is almost invisible when it sits still. Depth 6 m.

ED ROBINSON

carapace or the appendage and develop into an adult or give birth to a small colony. Thus, the camouflage of majids is often a result of active decorating by the crab and the presence of various epibiontic organisms.

Some crabs like *Cyclocoeloma tuberculata* shown on this page seem to prefer larger objects such as pieces of solitary corals or overgrown shells. Of all majid crabs of the Indo-Pacific, *Camposcia retusa* has probably the most perfect camouflage. This crab is virtually covered by pebbles and detritus, not only on the carapace, but also on the chelipeds and legs. The evolution of decorating behaviour is a fascinating topic of research. The most primitive condition is found today in crabs simply picking on detritus, as observed in the members of the family Latreillidae. This behaviour evolved further when some crabs started to place small animals and weeds on their front legs and claws. The purpose was probably the storage of portable and fresh food! Finally, this condition advanced in many species to decorating for the purpose of disguise.

A last word should be said about the hermit crabs (Paguridae and related families), which are not true crabs but belong to the section Anomura. They fall more or less into the category of allocryptism, since their mobile homes (usually empty gastropod shells) offer both concealment and physical protection.

This spider crab *(Cyclocoeloma tuberculata)* sitting between urchin spines is covered by two large and curious plates, probably pieces of solitary corals with tentacles. Sulawesi.

HELMUT DEBELIUS

The spider crab *Camposcia retusa* completely concealed by pebbles and detritus. Philippines.

PETER NG

Las pequeñas patas posteriores de este cangrejo de Tailandia forman pinzas para sujetar los detritos de camuflaje.

Diagramma labiosum **Sumatra**

Slate sweetlips

Le: To 100 cm. Di: Entire area. De: To 30 m. Ge: Juveniles singly, forming small to large groups when loosing the striped pattern. Coastal and inner reefs. Adults usually seen hovering in small groups above reefs or in caves during the day.

Plectorhinchus chaetodonoides **Sangihe I., Indonesia**

Harlequin sweetlips

Le: To 60 cm. Di: Entire area. De: 3-50 m. Ge: Large juveniles and adults may share large ledges along drop-offs. Small juveniles in protected bays and lagoons. Juvenile (below) brown with large white patches, intermediates with large black spots. Large adults have small spots over most of body and head.

Plectorhinchus picus **Flores, Indonesia**

Magpie sweetlips

Le: To 90 cm. Di: Indo-Pacific, entire area. De: 10-30 m. Ge: Seen near outer reefs with clean sand and large coral plates for sheltering during the day. Shown are young adults, some juvenile pattern still being evident. Below: Juvenile.

Oriental sweetlips

Le: To 50 cm. Di: Entire area. De: 3-50 m. Ge: Juveniles singly among boulders or low coral reefs in shallow depths. Adults form schools on current-prone coastal and inner reefs along slopes or the bottom of drop-offs. Below: Juvenile.

Plectorhinchus vittatus **Bali, Indonesia**

Yellow-ribbon sweetlips

Le: To 50 cm. Di: Entire area. De: 3-50 m. Ge: Juveniles singly on deep coastal rubble slopes. Adults in groups, also mixed with similar species. Number of stripes increases with age. On deeper part of current-prone reefs. Below: **Lined sweetlips** *P. lessoni*, to 40 cm, entire area, 1-50 m.

Plectorhinchus polytaenia **Nabucco I., Indonesia**

Oblique-banded sweetlips

Le: To 60 cm. Di: Entire area. De: 3-50 m. Ge: Juveniles singly in lagoons or on coral reef slopes. Adults on outer reefs along drop-offs, in some areas also forming large schools. Below: **Orange-lined sweetlips** *P. celebicus*, to 40 cm, entire area, 3-20 m.

Plectorhinchus lineatus **Okinawa, Japan**

The **Emperors** are medium-sized fishes, comprising 5 genera and 39 species, distributed in the Indo-Pacific, except one that occurs in the East Pacific. The largest genus is *Lethrinus* many species of which are encountered in the shallows. These fishes are mainly found on or near tropical reefs and only a few range into warm-temperate waters. They are bottom dwellers, carnivorous, feeding on a variety of invertebrates and small fishes. Some species feed only at night, resting on reefs during the day, but others feed during both periods and may alternate between open sand and reef. Most species, however, feed on or near reefs, either singly or in schools. In some areas these excellent food fishes can be ciguatoxic. Eggs are pelagic, small, less than 1 mm in diameter and produced in vast numbers. Larvae hatch at less than 2 mm in length, postlarvae are about 30 mm long. Many species settle in seagrass beds, are camouflaged and can change colour quickly. Most juveniles look very different from adults and some have different intermediate colour phases.

Two of the 5 genera (65 spp.) of the family **Nemipteridae (Spinecheeks, Threadfin breams,** p.124) include primarily deep water species. *Nemipterus* is the larges genus with 26 species, most encountered is *Scolopsis* with 18 species. The species included here are benthic feeders mainly active during the day. At night they rest on a substrate of rocks and corals. They feed on small crustaceans and worms (some also on small fishes, which are located with their excellent eyesight) and take mouths full of sand to sift it for prey.

Gnathodentex aureolineatus **Pulau Seribu, Indonesia**

Yellowspot emperor

Le: To 30 cm. Di: Entire area. De: 2-30 m. Ge: A common species of shallow coral reef areas. Often roaming in small groups (left) directly on the reef, rarely in huge schools.

Below: Single specimens prefer to hide in cavities of steep reef drop-offs.

Monotaxis grandoculis **Cebu, Philippines**

Big-eye emperor

Le: To 60 cm. Di: Entire area. De: 1-60 m. Ge: Small juveniles singly on rubble patches in lagoons or on reef crests from clear coastal to outer reefs. Adults congregate during the day (left) near drop-offs or large caves, often in large numbers, moving out at night to feed on the substrate in deep water.

Bluescale emperor
Le: To 80 cm. Di: Entire area. De: 2-50 m. Ge: Very common, often in small groups over corals and seagrass, also in huge schools over large rubble or sand areas. Below: **Slender emperor** *L. variegatus*, 20 cm, entire area, 1-30 m. Coastal algae reef, young in seagrass. More slender than other spp.

Lethrinus nebulosus **Alor I., Indonesia**

Longnosed emperor
Le: To 100 cm. Di: Entire area. De: 5-35 m. Ge: No distinct body markings but body clearly elongate. Swift, eats benthic fishes, molluscs, echinoderms. Below: Similiar **Small-tooth emperor** *L. microdon*, 70 cm, entire area, 3-80 m. On sand slopes near deep water, with mottled pattern while feeding.

Lethrinus olivaceus **Flores, Indonesia**

Blue-spotted large-eye bream
Le: To 40 cm. Di: Entire area. De: 20-50 m. Ge: On sand slopes, feeding on benthic invertebrates. Swims well above bottom, stopping to search bottom for food, diving quickly at prey. Below: **Striped emperor** *L. ornatus*, to 40 cm, entire area, 1-30 m.

Gymnocranius microdon **Bali, Indonesia**

Scolopsis bilineatus **Bali, Indonesia**

Two-line spinecheek
Le: To 23 cm. Di: Entire area. De: 1-25 m. Ge: Often seen hovering close over sand patches close to corals. It will move a short distance then hover again. Feeds mainly on invertebrates in the sand. Easy to approach. Below: **Black & white spinecheek** *S. lineatus*, to 23 cm, entire area, 1-25 m.

Scolopsis xenochrous **Taiwan**

Blue-stripe spinecheek
Le: To 20 cm. Di: Entire area. De: 5-50 m. Ge: Adults benthic in small groups on deeper parts of coastal rubble slopes, eat small crustaceans. Juv. with blue and black lines along sides, singly, shallower than adults. Below: **Bald-spot spinecheek** *S. temporalis*, to 30 cm, Sulawesi, Flores, PNG, 10-30 m, rare.

Scolopsis vosmeri **Gulf of Thailand**

Whitecheek spinecheek

Le: To 25 cm. Di: Entire area, not around ocanic islands. De: 5-45 m. Ge: In inner reefs and lagoons, also in estuaries. Singly or in pairs rooting for bottom-dwelling invertebrates. Below: **Lattice spinecheek** *S. monogramma*, to 30 cm, widespread tropical West Pacific, 3-50 m.

Golden spinecheek

Le: To 25 cm. Di: Sumatra and Java. De: 10-50 m. Ge: Juveniles coastal on sand flats between reefs. Adults usually in loose groups on sand and rubble slopes in depths of about 20 m or more. Below: **Yellow-tail spinecheek** *S. affinis,* to 25 cm, widespread tropical West Pacific, 20-60 m.

Scolopsis aurata **Bali, Indonesia**

Striped whiptail

Length: To 20 cm.
Distribution: Indonesia, Philippines and New Guinea.
Depth: 3-15 m.
General: A solitary species usually seen roaming over the reef tops, mainly in coastal and inner reef zones, but it also enters lagoons and harbours. Juveniles with darker and more distinct stripes, they are usually found in loose group on very shallow coastal sand flats.

Pentapodus trivittatus **Mindoro, Philippines**

Butterfly whiptail

Length: To 20 cm.
Distribution: Indonesia to Philippines, Singapore and south China Sea.
Depth: 10-100 m.
General: Coastal sandy areas between reefs, usually in 15 m or more. Below: The similarly widespread *P. caninus*.

Pentapodus setosus **Flores, Indonesia**

All family members are easily identified by the presence of a pair of barbels on the chin. The family comprises 6 genera and about 35 species with a global distribution in tropical and subtropical seas. They typically have elongated bodies with large scales, two separate dorsal fins and a forked tail. All fins are pointed or angular. Except for their coloration, many species of the same genus are difficult to separate on morphological features alone.

Goatfishes are benthic feeders and use their strong barbels to dig and sense prey in the substrate. The larger species search for food singly or have a few companions but often these are individuals belonging to non-related species such as wrasses and even carangids which keep an eye on the digging goatfish in case something tasty appears. Some of the smaller species and juveniles of the larger species school and may travel along the substrate while feeding at a rapid rate. On muddy ground this can create a large column of turbid water which is left behind on their path. They are pelagic spawners and produce tiny eggs, less than 1 mm in diameter, the pelagic larvae get quite large and reach 50 mm before settling down.

Parupeneus multifasciatus **Layang I., Malaysia**

Banded goatfish

Length: To 35 cm.
Distribution: Indo-Pacific, throughout our area.
Depth: 1-140 m.
General: Observed almost everywhere on sand and rubble near reefs at all depths, singly or in small groups. Juveniles are similar to the adults but they are more slender as is typical for all genus members. Best identified by the black areas below the second dorsal fin and on the caudal peduncle.

Parupeneus macronema **Singapore**

Long-barbel goatfish

Length: To 30 cm.
Distribution: Indo-Pacific, throughout our area.
Depth: 1-25 m.
General: A coastal reef species, preferring rocky areas, and often found singly or in pairs. Like in most species of the genus the colour of the body can turn bright red and this species seems to do this more often than others. It is easily overlooked because it resembles the more common *P. barberinus* but the stripe is broader and followed by a pale area which is uniform with the rest of the body.

Yellow-saddle goatfish
Le: To 40 cm. Di: Indo-Pacific, throughout our area. De: 3-100 m. Ge: Various habitats from shallow coastal to deep offshore. Adults singly or in pairs, also in small groups. Juveniles singly but usually mix with other species such as wrasses, especially *Thalassoma* spp. Below: Yellow variant.

Parupeneus cyclostomus **Bohol I., Philippines**

Dash-and-dot goatfish
Le: To 40 cm. Di: Entire area. De: 5-100 m. Ge: Coastal reefs and lagoons, singly or in small numbers, often followed by wrasses. Juveniles whitish with a black stripe. Adults with or without yellow above the stripe. Below: **Small-spot goatfish** *P. heptacanthus*, to 35 cm, entire area, 3-60 m.

Parupeneus barberinus **Flores, Indonesia**

Half-and-half goatfish
Le: To 30 cm, usually to 20 cm. Di: Entire area. De: 3-40 m. Ge: Smallest of genus, easy to identify by coloration. Solitary, juveniles in shallow bays among algae, adults deeper on rubble slopes with mixed coral and algae. Below: **Double-bar goatfish** *P. bifasciatus*, to 30 cm, entire area, 1-80 m.

Parupeneus barberinoides **Flores, Indonesia**

Mulloidichthys vanicolensis Derawan I., Indonesia

Yellow-stripe goatfish
Le: To 35 cm. Di: Entire area. Depth: 5-100 m. General: A schooling species, often in large numbers. Without spot. 2nd below: **Square-spot goatfish** *M. flavolineatus*, to 40 cm, entire area, 1-30 m. With spot.

Upeneus tragula Bali, Indonesia

Bar-tail goatfish

Length: To 30 cm.
Distribution: Indo-Pacific, entire area.
Depth: 3-30 m.
General: Coastal reef slopes and lagoons. Adults in small groups or singly. This species can change colour rapidly and when swimming it turns plain white with a dark mid-lateral stripe. There are several similar species, some still undescribed. Most of the smaller ones occur in large schools in muddy and coastal areas, and in addition there are some which prefer deep water and are only known from trawls.

Upeneus vittatus Flores, Indonesia

Striped goatfish

Length: To 25 cm.
Distribution: Indo-Pacific, entire area.
Depth: 5-100 m.
General: A very common species schooling in muddy coastal areas on slopes and deep sand flats. One of several similar species which feed when numerous individuals swimming side by side form a long line which moves at a rapid pace and usually creates murky water in the process.

Greenback bullseye
Le: To 12 cm. Di: Entire area. De: 3-20 m. Ge: Secretive by day, in large numbers amongst large boulders in coastal areas, or large caves. Best identified by colour and straight lateral line with short curve at origin. Adults of the about 20 spp. of this family usually live secretive in the reef or inside caves.

Pempheris vanicolensis **Alor I., Indonesia**

Black-margin bullseye
Le: To 16 cm. Di: Entire area. De: 3-20 m. Ge: During the day found along drop-offs and slopes, in small caves or under overhangs, often singly on coastal to outer reefs. With a straight lateral line that has a long curve at its origin. Below: **Black-base bullseye** *P. adusta*, 17 cm, entire area, 9-30 m.

Pempheris mangula **Bali, Indonesia**

Yellow sweeper
Le: To 10 cm. Di: Entire area. De: 10-50 m. Ge: A nocturnal species, forming large dense schools, often filling entire large caves along drop-offs, or deep around isolated bommies on sand far from main reefs. Disperse at night to feed in open water on zooplankton. School at right, close-up below.

Parapriacanthus ransonneti **Sangihe I., Indonesia**

ISLAND OF THE TURTLES

Within a period of 5 minutes I had seen two sharks swim by, a great barracuda *(Sphyraena barracuda)*, a huge bumphead wrasse *(Cheilinus undulatus)*, weighing perhaps 25 kg, three sea turtles *(Chelonia mydas)*, a swirling school of bigeye jacks *(Caranx sexfasciatus)*, a school of chub *(Kyphosus* sp.), and assorted other fish. On the shallow reef above the drop-off I could see more than two dozen species of hard corals, in addition to soft corals, anemones with their symbiotic clownfish *(Amphiprion* sp.), etc. At the lip of the drop-off, the reef plunged vertically from a depth of 2 m to a depth of 1,000 m. It would have been one of the most spectacular dives of my life. Except that I wasn't diving. Underwater photographer Doug Perrine narrates.

PHOTOS: DOUG PERRINE

Green sea turtle *(Chelonia mydas)* and Bigeye jacks *(Caranx sexfasciatus)* in a daytime "refuging" school at Sipadan Island, Borneo, Malaysia.

No, this wasn't a dive, but a surface interval. I was standing on the deck of the Drop-Off Cafe, sipping a hot drink, and soaking up the sunshine between dives. Not that I had become seriously chilled in the 29°Celsius water, but you do get acclimated to the tropical heat, and I had not worn an exposure suit since arriving on Sipadan. The Drop-Off Cafe is actually not on the island of Sipadan, but quite literally on the drop-off. It is connected to the beach in front of the Borneo Divers lodge by a short wooden pier. Because of the unique geography of the island, it was quite easy to erect a structure from which, on a calm day (which is to say an average day) it is possible to gaze down over the edge of the railing and observe both the life on an idyllic tropical reef, and the pelagic species that cruise in the deep blue water just beyond the edge of the drop-off.

Of course this makes for some fairly spectacular shore-diving as well. It's easy to just grab a tank (and a buddy) at the dive shop, stroll across the beach, kick across the reef to the edge of the drop-off, and float down over the lip of the wall into that netherworld where an expanse of endless blue is bounded on one side only by a surrealist canvas where every colour of the rainbow leaps out wherever the probe of a dive light beam is applied. An efficient diver could probably make it from the door of his cabin to a depth of 25 m in just over 5 minutes. The wall is vertical in some areas, and actually undercut in others. The most violent explosions of colour occur in the undercut areas, which are covered with orange cup corals *(Tubastraea* sp.) and other encrusting organisms. A variety of exotic nudibranchs forage and lay their eggs among the encrusting corals and other "foliage." Here and there, cleaner wrasses *(Labroides dimidiatus)* have set up their parasite-removal stations, and a parade of larger fishes queue up to be serviced.

At night, dozens of giant bumphead parrotfish *(Bolbometopon muricatum)* come to this particular section of the wall to sleep. Every indentation in the wall contains one or more of these massive emerald-hued fish, some over a metre in length and weighing up to 40 kg or more. The crimson colour of the forehead projection on the large males is brilliantly visible with a torch at close range at night. A short swim down the wall from the entry point is a large cavern, at a depth of about 20 m. Deep inside are the skeletons of sea turtles which swam inside to sleep at night, and, awaking to go to the surface for a breath of air, could not find their way back to the entrance and drowned. It is wise to remember that sea turtles are much better at this sort of navigation than we are, and that the same fate could easily befall a scuba diver (and has, on a few unhappy occasions). Signs mark the limit of the cav-

Mating Green sea turtles are joined by another male (underneath) for a *menage a trois,* while a third male approaches behind.

ern that can be safely explored without a guide. As the skeletons are farther inside, it is necessary to arrange with a divemaster for a tour to view them. However, a guide is not necessary for the most spectacular show on the island, which occurs right at the entrance to the cavern, but only at night.

At night, however, the ledge is a perfect place for a buddy team to settle down on their knees, turn out their dive lights, and wait. The few minutes that are required to accommodate our human vision to darkness are about the same amount of time necessary for flashlight fish *(Photoblepharon* sp.) to overcome the shock of seeing divers' torches at the front of their cavern, and to resume their courtship activities. Peering into the darkness at the back of the cave, the divers can dimly see clusters of greenish lights zipping around erratically like UFOs in the night sky of an old science fiction movie. The light clusters zigzag closer and closer until they reach the lip of the cavern roof, just above the kneeling divers. Here the clusters crash into each other forming explosions of phosphorescent light (as gametes are presumably being released to spawn future generations). If a torch is turned on, the fish go streaming towards the back of the cavern so fast that it is rare to catch even a glimpse of the tiny fish that put on this stupendous light show.

With such tremendous entertainment available just in front of the lodge, you might wonder why one would bother with boarding a dive boat, just to motor 5-10 minutes away to some other part of the same reef circling the tiny (four hectare) island that is Sipadan. To be honest, I had exactly the same thoughts before coming. I had figured on spending almost all of my time shore diving. That only lasted until I did

Skeleton of a Green sea turtle which drowned in "Turtle Cavern" at Sipadan Island.

A Green sea turtle hatchling swims at the surface after reaching the ocean.

my first boat dive. Even on a small reef, all parts of the reef are not equal. Different sections of the reef are prone to different currents, which attract different collections of fish and other life. The currents range from non-existent to violent and irresistible, and can accelerate or even completely change direction during the course of a dive. Additionally, the width of the shallow reef shelf varies from extremely narrow in some places to quite broad around much of the island. Therefore, shore diving is impractical except at this one short section of the coastline. Elsewhere around the island, there are daily spectacular aggregations of marine life that can only be accessed by boat.

There are spots where, at slack tide, schools of barracuda *(Sphyraena quenie)* and bigeye jacks *(Caranx sexfasciatus)* form massive whirling gyres of fish. Under the right conditions, boat drivers can spot these schools from the surface, and drop off divers in a position where they can descend straight down into the centre of the gyre. If the divers are careful, and stay away from the spinning walls of the cylinder of fish, several can enter the "inner sanctum" at once.

Of course, that's not what usually happens. Usually too many divers all try to get too close to the school of fish at once, and cause the school to either break up and move. Even if all the divers in one group are cautious and well behaved, there is a good chance that another group of divers will come along and all try to join in at once. If there's only a single disturbance, the school can move down the reef and re-form, but if another group of divers comes along the fish have to keep moving or go deep, and may suffer from stress, lack of rest, increased predation, higher parasite loads due to interruption of cleaning services, etc. It happens with the jacks, the barracuda, the bumphead parrotfish, the great schools of batfish *(Platax* sp.) which flutter across the reef like heavenly angels, and the turtles. The turtles are fairly implacable, but if given too much attention, will eventually leave their resting places and move off, expending needed energy in the process. Did I not mention the turtles before? That's because there are so many of them that you almost cease noticing them after a while. But more about turtles later.

The point is not that divers are bad for the reef or the marine life. Indeed, the presence of divers and dive resorts is probably the main thing that has saved Sipadan from being pillaged and destroyed by unscrupulous fishermen, like so much of the surrounding area. The presence of so many divers is also the reason why so many of the fish are "tame" and easily observed. They have had no choice but to adapt themselves to the new presence in their world. But the fact is that fish and other wildlife can only adapt to a certain extent. The fish, the corals, and even the island itself can only handle a certain level of human activity. Nearly everyone who has been to Sipadan in the last ten years agrees that a sustainable level of human impact was surpassed long ago, resulting in a degradation of water quality, erosion of the beaches, and various hazards to the wildlife. Five dive resorts host up to 200 divers at a time, plus as many more may come over to dive each day from resorts on the nearby island of Mabul.

The regional Malaysian government has determined that, in order to protect and nurture the natural resources of Sipadan, no more than 80 divers should be accommodated on the island at a time, with an additional small number allowed to come over from Mabul. Therefore, each of the dive resorts has been given a quota of divers allowed. However, this edict has not yet been effectively enforced at the time of writing.

Does this mean that you should not risk booking a trip to Sipadan? Not at all. The diving, among the most fantastic I have ever experienced, should get even better once the number of divers is brought down to a reasonable level. The key to avoiding problems would be to ensure that the resort you book with is limiting their bookings to the permitted number. Otherwise, you could experience a last-minute unexpected cancellation. Moreover, it is morally responsible to do business with the operations that comply with conservation measures.

In fact, the diving is already better than it has been in recent years due to the enforcement of laws banning dyna-

Hawksbill turtle *(Eretmochelys imbricata)*.

mite fishing in the surrounding area. Few sticks of dynamite have been lobbed onto the reefs of Sipadan recently, but the noise of explosions on nearby reefs could range from disconcerting to downright scary. During my ten-day stay on Sipadan, I didn't hear a single fish bomb go off. There is some evidence that parts of the reef may have been bombed in the past, but to a large extent, the island has been spared the depredations that have been visited upon many other reefs in the vicinity. This may have been partly due to its remote location, partly to an old superstition about a giant octopus that supposedly ate fishermen, and perhaps partly due to laws that have protected the island as a sanctuary for birds and other wildlife even before it became a Mecca for divers and was designated as a marine reserve. The island shelters 39 species of birds, including the endangered Nicobar pigeon and the strange and rare megapode bird. There are also monitor lizards well over a metre long and large coconut crabs (nearly extinct from the rest of the region), and the island is a major nesting spot for green sea turtles.

Lack of fishing pressure cannot totally explain the amazing concentration of life found around this tiny island. Some credit must also be given to its location and geography. The island was formed by the steady accumulation of coral growth on the crown of an undersea volcano which became extinct millions of years ago, and has been sinking and eroding as the coral has grown upwards. Now a cross sectional view would show an island shaped like the head of a pin, sitting on a narrow shaft that extends down 1,000 m to a sea bottom next to an even deeper trench. It is separated from the shallow continental shelf by a distance of about 10 km, and from the large island of Borneo by several times that distance.

The island is crowned by tropical forest that has been left largely intact by the resort development. There is a white sand beach all the way around the island (although it is so narrow in a few places that passage is difficult at high tide). It takes about 20 minutes to walk around the island. Each year, hundreds of green sea turtles crawl onto this beach to lay their eggs. During my stay, two crawled up and dug their nests right in front of the resort. Prior to nesting, the turtles mate. At the peak of the season (July-October), the waters around Sipadan become a sea of love, with coupling turtles a frequent sight. Wading out from the beach for a night dive, we encountered a pair of turtles just in front of the beach in only 2 m of water, coupling blissfully in the light of the full moon.

At night, there seems to be a turtle asleep in every nook and cranny of the reef. By day there are turtles resting in most of these spots, turtles scratching their shells on the coral to remove barnacles, turtles waiting at cleaning stations for surgeonfish to graze the algae off their shells, male turtles with great long tails swimming around looking for ripe females, turtles floating at the surface breathing, turtles swimming around aimlessly, turtles just hanging around and getting in the way. In addition to the hundreds of green turtles, there are a smaller number of hawksbill turtles *(Eretmochelys imbricata)*, which feed heartily on the sponges and leather corals, and rummage through the coral rubble looking for other titbits, attracting a coterie of reef fish hoping to snag the leftovers.

Sipadan more than earns its reputation as the "island of the turtles." But it is much more than that. It is a wilderness paradise where the haunting cry of the megapode bird echoes through the jungle at dawn, 1.5-metre lizards wander into buildings unannounced, and divers vanish inside schools of fish. It is a speck of land in a deep blue sea, so small that each dive site is visited again and again over the course of a week, yet so alive that no dive site is ever the same twice.

The small but interesting family Ephippidae is represented by 7 genera of which only *Platax* spp. are commonly seen in our area (the others are found in deep water). All known five species of *Platax* are included here. The fifth species, *P. boersi,* was recognised as being distinct only recently, earlier it was confused with the other species. The various species are known as batfishes and spadefishes which relates to their shape. In general the body is almost circular and the rest of the profile is added by the shape of the fins. In some juveniles the dorsal and anal fin are extremely tall and deep like wings which proportionally decrease in size with age and most adults become more 'spade'-shaped. Small juveniles are often very attractive and sold in the aquarium-fish trade. Unfortunately they grow quickly and to a large size, and most people will not be able to support the adult fish. However, they make great pets and if space is not a problem they are highly recommended.

Adults can be found from coastal lagoons to outer reefs where they may form large schools, and have been trawled to depths of 500 m. They are thought to be pelagic spawners and their larvae transform into the familiar, sometimes very tall juveniles when about 20 mm long. Juveniles are either pelagic or benthic, depending on the species, usually seeking quiet waters in coastal bays and estuaries. Diet consist of algae and a variety of invertebrates such as jellies or other plankton. Juvenile *Platax* have interesting early stages. The tiny juveniles mimic floating leaves, crinoids and one even a nasty tasting flatworm. Adult form schools to travel through open water in tight schools that take on shapes that from a distance look like large sea creatures. One school seen from a distance in clear water off an outer-reef wall was mistaken for a sunfish!

Platax orbicularis **Banda Sea, Indonesia**

Round batfish

Length: To 50 cm.
Distribution: Widespread tropical Indo-Pacific, entire area.
Depth: 1-40 m.
General: Small juveniles float at the surface with the current or wind-driven land debris, in particular orange-brown leaves. They settle in lagoons or along beaches on sand in small schools (left and facing page, photo from Sipadan Island, Malaysia). Adults pair or school in different areas and are usually found in deeper offshore areas. Bottom small photo: Juvenile.

Platax orbicularis **Derawan I., Indonesia**

Tall-fin batfish

Length: To 60 cm.
Distribution: Very widespread in tropical and sub-tropical Indo-Pacific, entire area.
Depth: 1-50+ m.
General: This is the most common batfish species encountered by divers in Asian Pacific waters. Large adults occur in shelterd bays as well as deep offshore. They are seen singly or in schools, often in the vicinity of shipwrecks, quickly adapting to visiting divers and often rising to the surface to meet them when the divers start their descend.

Small juveniles can be pelagic and form schools under large Sargassum rafts that become adrift after the wet season. A typical juvenile is shown at the left. The beautiful photo below shows a group of subadult Tailfin batfish hovering above a healthy coral reef for which the South East Asian region is well known.

Platax teira **Flores, Indonesia**

Tukang Besi, Indonesia

Platax teira **Okinawa, Japan**

Adult Tailfin batfish have a diagnostic black blotch posteriorly at the lower end of the dark band behind the head. The blotch is less obvious when the band is dark but it is the best character for distinction from the similarly looking adults of the species on the following pages. Left: Adult *Platax teira* in front of a wreck.

Shaded batfish

Length: To 35 cm.
Distribution: Tropical West Pacific, entire area.
Depth: 2-50 m.
General: Postlarvae settle on coastal protected reefs or in mangroves. Juveniles are often seen under jetties but also move to deep water where they inhabit caves. Adults are usually found living singly or in pairs from shallow coastal to deep outer reefs, and are often also seen in shipwrecks. Small juveniles are black with bright orange-surround outline. This gives them the appearance of a nasty tasting flatworm.

Diving for more than 10 years in South east Asian waters the author saw many schooling batfish of other species, but never had come across schooling *Platax pinnatus*. It was a calm day near a drop-off at the Lesser Sunda island of Flores when he had almost finished his dive hovering in the reef at a depth of 10 metres. The water was very clear and visibility was excellent. Looking down the wall, he suddenly noticed a school of about 20-30 adult batfish. He also could see the typical pointed snout and realised that these fish were Shaded batfishes. Immediately he dropped to a depth of 45 metres and there was able to obtain his first photograph of an aggregation of this species. In contrast to other species, the fish were very shy and left the scene into even deeper waters almost instantly after the first flash.

Platax pinnatus

All photos: Lesser Sunda Islands, Indonesia

Platax pinnatus

Platax batavianus **Sumatra**

Zebra batfish

Le: To 65 cm. Di: Tropical West Pacific, entire area. De: 15-60 m. Ge: Large, usually singly or in small loose groups in deep lagoons or on sandy flats between reefs. Juveniles associate with crinoids. Adults become humped, they live either singly or in small groups. The least observed batfish species. Juvenile below.

Platax boersi **Sipadan I., Malaysia**

Boers batfish

Le: To 40 cm. Di: Indonesia, New Guinea and Philippines. De: 1-50 m. Ge: Adults usually in large schools along drop-offs, including outer reefs, but also observed singly in coastal waters. Because of the shape usually confused with the round batfish, *P. orbicularis.* Small juveniles are easily confused with *P. teira* and best separated by the shape of the median fins. Juvenile below.

DRUMMERS — KYPHOSIDAE

Brassy drummer
Le: To 40 cm. Di: Entire area. De: 1-20 m. Ge: Coastal reef flats, slopes and lagoons. Swims in midwater, often singly or in small groups. Very shallow to outer reefs in exposed areas. Below: Similar **Snubnose drummer** *K. cinerascens*, to 50 cm, entire area, 1-20 m.

Kyphosus vaigiensis **Pulau Seribu, Indonesia**

SILVER BATFISHES — MONODACTYLIDAE

Silver batfish
Le: To 20 cm. Di: Entire area. De: 1-10 m. Ge: Colour dark in juveniles and silver in adults. Occurs mainly in large coastal estuaries in silty habitats, often in large schools around break-waters or under jetties. Juveniles in brackish water, sometimes moving into freshwater. Adults move from estuaries to coastal reefs. The Silver batfishes are a small family with three genera and five species, only *Monodactylus* occurs in our area. They have tiny ventral fins which become rudimentary in adults.

Monodactylus argenteus **Palau**

SCATS — SCATOPHAGIDAE

Spotted scat
Le: To 35 cm. Di: Entire area. De: 1-10 m. Ge: Variable with age. Small juveniles black with red, becoming spotted, first forming bands, and decreasing to small spots in adults. In coastal estuaries and river entrances, mainly in brackish water, including settling juveniles. The Scats are a small family with two genera and 3-4 species. They have compressed bodies, similar to that of the butterflyfishes, but 4 spines in the anal fin. The fin spines can inflict painful wounds and are thought to be venomous.

Scatophagus argus **Sumatra**

BUTTERFLYFISHES CHAETODONTIDAE

This large family of colourful fishes is popular with divers and aquarists. It comprises 10 genera with about 120 species which mostly inhabit coral reefs, but a few have adapted to warm temperate zones, and some live in deep water. The majority of species occur in the tropical Indo-Pacific with the largest concentration on Indonesian coral reefs, only 4 are known from the east Pacific and 12 from Atlantic waters. Most species are placed in the genus *Chaetodon* (114 spp. in 13 subgenera), *Heniochus* has 8 spp. and the rest are monotypic or with few spp. each.

Most butterflyfishes live in pairs in shallow depths on coral reefs, the species of *Hemitaurichthys* live in schools. Some species live in pairs in one region and school in another. Their diet consists of small invertebrates which are usually picked from the substrate, others feed on zooplankton. They produce tiny spherical pelagic eggs less than 1 mm in diameter. Their larvae have a bony head-armour, often with serrated spines, this stage is known as 'tholichthys'. The post-larvae can be as small as 10 mm long and settle in corals and rocks. Most species are very secretive.

Hemitaurichthys thompsoni **Ogasawara Is., Japan**

Thompson's pyramid butterflyfish

Length: To 18 cm.
Distribution: Oceanic islands in the eastern part of our area.
Depth: 10-300 m.
Ge: Usually on deep outer reefs, also shallow in lagoons and coastal reefs near deep water. Feeds on plankton.

Hemitaurichthys polylepis **Bunaken I., Indonesia**

Schooling pyramid butterflyfish

Le: To 15 cm. Di: Entire area. De: 5-60 m. Ge: In groups (left) along current-swept drop-offs in coastal as well as outer reef zones, where they feed along the edge on plankton drifting past. See also facing page.

Chaetodon argentatus **Yonaguni I., Japan**

Asian butterflyfish
Le: To 20 cm. Di: Philippines, Taiwan, S-China to S-Japan. De: 5-20 m. Ge: In pairs on rocky and coral reefs. Below: **Black-barred butterflyfish** *C. burgessi,* to 14 cm, Indonesia (Flores, Sulawesi) to Palau, 20-75 m. On outer reef drop-offs, usually in 40+ m. Common, occurring only in pairs.

Chaetodon xanthurus **Sangihe I., Indonesia**

Cross-hatch butterflyfish
Le: To 10 cm. Di: Indonesia, Philippines. De: 6-50 m. Ge: Adults pair along deep outer reef slopes and drop-offs. Swims over large areas in search for food, stopping for closer inspection. Juveniles singly, secretive. Below: **Lemon butterflyfish** *C. citrinellus*, to 11 cm, entire area, 1-36 m.

Chaetodon ocellicaudus **Sulawesi**

Tail-spot butterflyfish
Le: To 14 cm. Di: Indonesia to Philippines. De: 5-50 m. Ge: Singly or in pairs, usually on steep slopes and drop-offs of coastal to outer reefs. The distinct black spot on the caudal peduncle separates it from the similar *C. melannotus* below, which in the juvenile stage is even more pronounced.

Eye-patch butterflyfish
Le: To 20 cm. Di: Entire area. Java to southern Japan. De: 3-30 m. Ge: Coastal reef tops and slopes in hydroid and soft coral areas. Usually in pairs, also in small groups, often swimming in midwater and showing little or no activity during the day. Juveniles similar to adults and singly on reefs.

Chaetodon adiergastos Layang I., Malaysia

Spot-banded butterflyfish
Le: To 10 cm. Di: Entire area. De: 6-45 m. Gel: Usually in pairs along steep outer reef drop-offs. Active, often shy, eats invertebrates off rock. Juveniles very secretive. Bands nearly vertical. Below: Very similar *C. pelewensis* (below, bands diagonal), in same habitat, hybrids with mixed patterns.

Chaetodon punctatofasciatus Sulawesi

Reticulated butterflyfish
C. reticulatus
Le: To 20 cm. Di: Northern Indonesia to Philippines, northern New Guinea to New Caledonia. De: 1-30 m. Ge: An oceanic species with a preference for clear water reefs.

Teardrop butterflyfish
C. unimaculatus
Le: To 20 cm. Di: West Pacific, entire area. De: 1-30 m. Ge: In our area nearly always in pairs, uncommon in central Indonesian waters and more abundant at oceanic locations. Further west replaced by yellow variant.

Chaetodon reticulatus & C. unimaculatus Ambon, Indonesia

Chaetodon ornatissimus **Kerama I., Japan**

Ornate butterflyfish

Le: To 20 cm.
Di: Tropical West Pacific, entire area. De: 1-40 m.
Ge: Almost always in pairs swimming over reef crests near drop-offs. Feeds on corals, nibbling at the polyps. The similar, closely related *C. meyeri* is shown below.

Chaetodon trifascialis **Talaud I., Indonesia**

Chevroned butterflyfish

Le: To 14 cm. Di: Entire area. De: 2-20 m. Ge: Primarily with *Acropora* corals, singly or in pairs and often juveniles are present as well. Juveniles have the black on the tail more forward on body and fins. Found from coastal to outer reefs. Left a group of adults and juveniles, adult below.

Chaetodon plebeius **Bali, Indonesia**

Blue-dash butterflyfish

Le: To 12 cm. Di: From Entire area to southern Japan. De: 2-15 m. Ge: Singly or in pairs in clear water, feeding on coral polyps. Juveniles secretive between coral branches. Below: The similar, closely related **Andaman butterflyfish** *C. andamanensis,* to Sumatra.

Raccoon butterflyfish
Le: To 20 cm. Di: Entire area. De: 1-25 m. Ge: Juveniles on coastal reefs and in harbours, secretive in shallow areas, usually in 1-2 m depth. Adults in various habitats from coastal to outer reefs and commonly found on shipwrecks. They pair in Indonesia, but school at some oceanic localities.

Chaetodon lunula **Palau**

Pacific pinstriped butterflyfish
Le: To 12 cm. Di: Entire area. De: 3-20 m. Ge: In pairs or small groups in coastal coral-rich areas. Small juveniles hide amongst branches of small coral heads, mostly in lagoons. The **Indian pinstriped butterflyfish** *C. trifasciatus* (below), Indian Ocean to Bali.

Chaetodon lunulatus **Mabul I., Malaysia**

Günther's butterflyfish
Le: To 14 cm. Di: Celebes and PNG to S-Japan. De: 5-40+ m. Ge: Coral-rich outer reef slopes, rocky reefs. Below: **Yellow-dotted butterflyfish** *C. selene,* to 20 cm, entire area, 15-50 m. Deep coastal rubble and sand slopes with mixed reef patches. Usually in pairs close to the substrate.

Chaetodon guentheri **Ogasawara Is., Japan**

Chaetodon lineolatus **Nabucco I., Indonesia**

Lined butterflyfish
Le: To 100 cm. Di: Entire area. De: To 5-35 m. Ge: Largest sp. of the genus. In lagoons and seaward reefs, usually in pairs in coral-rich areas. Feeds on coral polyps and small anemones, also takes algae. Below: Similar **Pig-face butterflyfish** *C. oxycephalus,* to 30 cm, Indonesia to S-Japan, 10-40 m.

Chaetodon baronessa **Banda Sea, Indonesia**

Pacific triangular butterflyfish
Le: To 60 cm. Di: Entire area. De: 3-15 m. Ge: Adults singly or in pairs on reef crests and slopes with *Acropora* corals on which they feed as a major part of their diet. Below: **Vagabond butterflyfish** *C. vagabundus*, to 18 cm, Philippines, Malaysia, Thailand, 1-30 m. Lives in pairs.

Chaetodon octofasciatus **Palawan, Philippines**

Eight-banded butterflyfish
Le: To 12 cm. Di: Entire area. De: 3-20 m. Ge: Common, in sheltered areas close to shore in turbid water with corals on which it feeds. There are two colour forms, one yellow, another silver. Below: **Saddled Butterflyfish** *C. ephippium,* to 23 cm, entire area, 1-30 m.

Wrougt iron butterflyfish

Length: To 15 cm.
Distribution: Restricted to southern Japan and the Ryukyu, Ogasawara, and Izu Islands.
Depth: 2-10+ m.
General: Found in the clear shallow water of rocky reefs close to shore. In large groups which swim about to graze algae and small bottom-dwelling invertebrates from rocks. Rare except in southern Japan. To see this extravagant butterflyfish one definitely has to make a trip to Ogasawara Is. See also the photo on p.9.

Chaetodon daedalma **Ogasawara Is., Japan**

Dotted butterflyfish

Le: To 24cm. Di: Entire area. De: 3-50 m. Ge: Shy, usually in pairs on rich and clear coastal reef crests in channels, along slopes and drop-offs feeding on invertebrates from the bottom. Schools do migrate. Below: **Lattice butterflyfish** *C. rafflesi,* to 20 cm, entire area, 1-50 m. Singly or in pairs.

Chaetodon semeion **Sangihe I., Indonesia**

Orange-banded coralfish

Le: To 14 cm. Di: Entire area. De: 5-60 m. Ge: Singly or in pairs on deep coastal reef slopes. Often in cool upwelling areas. Juveniles similar but ocellus in dorsal fin much larger. In sponge areas, eats invertebrates. Below: Similar **Highfin coralfish** *C. altivelis*, to 15 cm, entire area, 3-20 m.

Coradion chrysozonus **Derawan I., Indonesia**

Heniochus diphreutes **Luzon, Philippines**

Schooling bannerfish

Length: To 20 cm.
Distribution: Indo-Pacific, entire area.
Depth: 1-100+ m.
General: A schooling species, thousands of adult individuals may be seen in current-swept channels to feed on passing zooplankton. Juveniles are more coastal and can be found on reef outcrops on mud or sand bottoms. They are active cleaners and sometimes approach divers.

Heniochus chrysostomus **Banda Sea, Indonesia**

Pennant bannerfish

Length: To 16 cm.
Distribution: Widespread in tropical West Pacific, including all of our area.
Depth: 1-40 m.
General: Adults usually in pairs along deep drop-offs but also found singly or in small groups from deep lagoons in clear coastal waters to outer reefs. Juveniles often live very shallow in lagoons with mixed coral and algae reefs. The diet consists of invertebrates such as coral polyps, polychaete worms and tiny crustaceans.

Heniochus singularius **Flores, Indonesia**

Singular bannerfish

Le: To 30 cm. Di: Indo-Pacific, entire area. De: 2-50 m. Ge: Adults in pairs on deep coastal and inner reef slopes, also on wrecks. Rather shy. Juv. usually singly in shallow lagoons but also deeper in small caves. Eats invertebrates and algae. Below: **Horned bannerfish** *H. varius*, to 20 cm, entire area, 2-30 m.

Beaked coralfish

Length: To 20 cm.
Distribution: Widespread in tropical West Pacific.
Depth: 1-15 m.
General: This butterflyfish species is seen singly or in pairs on coastal or inner reefs but also in large open estuaries. It feeds primarily on small invertebrates picked out of narrow crevices. The territorial species is abundant in some areas but rare or absent in central Indonesia.

Chelmon rostratus **Batangas, Philippines**

Long-nose butterflyfish
Le: To 22 cm. Di: Indo-Pacific, entire area. De: 3-30 m. Ge: Usually in pairs. Exposed reefs on crests and upper sections of drop-offs. Bottom small photo: Dark phase. Below: **Very-long-nose butterflyfish** *F. longirostris*, to 22 cm, entire area, 5-60 m. Adults usually in pairs, juveniles singly. Outer reefs along deeper part of drop-offs. Colour variations, in some areas entirely brown.

Forcipiger flavissimus **Singapore**

RESTLESS RUSTY ANGELFISH

The members of many reef fish genera are known to display more or less elaborate courting and spawning behaviour but not all of the species have been observed and/or photographed while doing so. While diving in tropical Japanese waters underwater photographer Yoji Ohkata was able to add the Rusty angelfish to the list of documented sexual behaviour of reef fishes. Translation by Hiroyuki Tanaka.

PHOTOS: YOJI OHKATA

During courtship the male Rusty angelfish displays brilliant colours of bright orange and neon blue.

Eleven species of the pygmy angelfish genus *Centropyge* are recorded from Japanese waters until now. Since they tend to lead a secretive life divers can hardly observe them for a long while in their natural habitat. However, I was able to observe a pair of the Rusty angelfish *Centropyge ferrugata* and their spawning behaviour at Amami-Ohshima Island in the Kagoshima Prefecture of southern Japan.

The species features a red belly and ranges from Tanabe Bay, Wakayama Prefecture, southward in the northwestern Pacific Ocean. It is commonly seen in the coral reefs of Okinawa, but seems more abundant on rubble areas than among corals, probably because this is where a lot of seaweeds grow which they feed on. At first, all individuals are females, later some of them change sex to become males. This is well-known for species of the genus *Centropyge*. The male has a harem and is leading the females. There is no big difference between male and female in most of the species, but *Centropyge ferrugata* is one of the exceptions. Males have neon blue markings at the posterior edge of the dorsal fin. Additionally females are considerably smaller than males. They always feed during the day and immediately take cover among rubble when danger approaches. Perhaps their relationship is closest during their reproductive stage. While they do not reproduce they are peaceful, even when two males have a close encounter.

The reproductive season starts at the beginning of summer and ends in fall. I observed them spawning at Amami-Ohshima several times in November, so spawning may depend on the water temperature which was 24°C in June and 25°C there in November 2000. Maybe their reproduction is busiest during the warmest September days. During the spawning season the coloration of some individuals becomes more vivid, especially the

The larger and more colourful male shakes his body in front of one of his females.

red colour. The blue areas of the males extend and become more conspicuous, which may attract females.

Spawning activities start in the evening and go on until both sexes concentrate on feeding. In the early evening the male becomes more active and patrols his territory. He gradually moves closer to one of his females who at first is not concerned at all. Then he approaches another female and this behaviour is repeated for a while. While dusk approaches he positions himself just above one of the harem members and shakes his body with fins outspread. At first she is not interested in his action and he soon moves to another female to do the same thing. After repeating his behaviour one of the females leaves the bottom and the male hurries towards her to display the same courtship behaviour. He nudges her belly with his snout, which is called nuzzling. Then they swim 40-50 cm towards the surface and release eggs and sperm when they are parallel to each other. Just after that they immediately return to the bottom. He displays the same behaviour to all of his females in turn during the evening but some females seem too 'nervous' and eventually do not spawn. They probably choose the time of the day when no predators are around. Nevertheless, such an interruption is still unusual for them. I wonder how they know about their timing of releasing eggs and sperm and why a male returns to the bottom first and his mate follows him.

Translator's Note: The 12th species of *Centropyge (C. colini)* in Japanese waters has just been documented and photographed at the Ogasawara Island.

The male swims up from the bottom above the female.

Now the male has engaged in nuzzling the side of the female.

Finally the pair rises above the ground and simultaneously releases sperm and eggs (visible as small white dots in the water).

This family comprises some of the most beautiful reef fishes, comparable to the butterflyfishes, however they are generally larger and more majestic. They are sought after by aquarists and highly prized. Over-exploiting species in some areas has led to banning imports of the entire family in many countries. They are readily distinguished from the closely related butterflyfishes by the large spine on the lower corner of the gill cover. The 7 genera with about 80 species fall into several distinct groups: The larger species, of which *Pomacanthus* spp. reach up to 35 cm; the small species known as pygmy angels, *Centropyge* spp., some reach only about 10 cm; and the planktivorous lyre-tailed *Genicanthus* with a distinct body shape. The larger species have the most dramatically differently coloured juveniles imaginable, not even slightly close to the adult, and *Genicanthus* has very different colour patterns between sexes. Juveniles of most species live secretively in reefs and are very territorial, particularly when facing their own kind but also towards close relatives.

Except for *Genicanthus* spp. they are benthic feeders, scraping and biting algae off coral rock, sponges or corals, also eating associated invertebrates, an activity which seems to last most of the day. These fishes are able to produce low-frequency, drum-like or thumping sounds, and when the large species do, the noise is so loud that it can startle a diver. Angelfishes produce small round pelagic eggs. Hatchlings are less than 2 mm long and some attain 25 mm before settling, however most species settle when 10 mm long.

Pomacanthus imperator **Mabul I., Malaysia**

Emperor angelfish
Le: To 38 cm. Di: Entire area. De: 6-60 m. Ge: Adults usually in pairs, in deep lagoons with bommies and at reef walls in caves, on coastal and outer reefs. Small juveniles (below) are found usually deep. Compare juveniles on these two pages, differing considerably from adults and each other.

Pomacanthus annularis **Sumatra**

Blue-ringed angelfish

Length: To 45 cm. Distribution: Indonesia to southern Japan. Depth: 1-60 m. General: A large and very impressive fish, commonly found under jetties, in harbours and in shipwrecks. Adults usually live in pairs at depths of 20 m or more. Juvenile below.

Six-banded angelfish

Le: To 46 cm. Di: Entire area. De: 1-60 m. Ge: A large coastal species, adults are usually seen in pairs. The juvenile (below) is best identified by the mainly white lines on the body and a plain pectoral fin.

Pomacanthus sexstriatus **Tukang Besi, Indonesia**

Blue-face angelfish

Length: To 36 cm.
Distribution: Entire area.
Depth: 5-45 m.
General: In reefs with plenty of coral growth but also with many channels and caves. Adults mostly singly, juveniles (below) secretive on shallow reefs.

Pomacanthus xanthometopon **Palau**

Majestic angelfish

Length: To 25 cm.
Distribution: Widespread in tropical West Pacific. Depth: 6-40 m. General: At outer reef drop-offs and in deep lagoons with rich coral growth. Juveniles (below) live in outer reef lagoons and are best identified by the blue pectoral fin.

Pomacanthus navarchus **Flores, Indonesia**

Chaetodontoplus septentrionalis **Ryukyu Is., Japan**

Blue-striped angelfish
Le: To 22 cm. Di: Tropical Japan, south to Viet Nam, Malaysia. De: 10-60 m. Ge: Common in Japan, ranges far into China Sea. Coastal rocky reefs along walls near ledges or caves. Juveniles secretive. Feeds on sponges and tunicates. Below: *C. chrysocephalus,* to 22 cm, entire area except at the oceanic islands, 15-25 m.

Chaetodontoplus melanosoma **Palawan, Philippines**

Black velvet angelfish
Le: To 20 cm. Di: Entire area except at the oceanic islands. De: 5-30 m. Ge: With colour variants, grey restricted to head in PNG. Solitary or in small groups. Below: **Black angelfish** *C. niger,* to 35 cm, southern Honshu, Japan, and Macclesfield Bank between Philippines and Viet Nam, 0-80 m. Very rare.

Chaetodontoplus mesoleucus **Sulawesi**

Vermiculate angelfish
Le: To 18 cm. Di: Tropical West Pacific from Indonesia to southern Japan. De: 6-30 m. Ge: Inhabits rich coral growth areas of coastal reefs and lagoons. Adults nearly always in pairs, swimming close together. Small juveniles singly and secretive in corals. Below: Colour variant.

Three-spot angelfish
Le: To 26 cm. Di: Entire area. De: 3-60 m. Ge: Coastal reef slopes on rich rubble ridges with soft corals, sponges. Adults in small groups where common, juveniles secretive on rubble, especially on deep isolated outcrops with sponges in 30+ m. Below: *A. griffisi,* to 25 cm, Sulawesi, Palau, PNG, 10-60 m.

Apolemichthys trimaculatus **Tioman I., Malaysia**

Herald's pygmy angelfish
Le: To 10 cm. Di: Entire area. De: 6-60 m. Ge: Usually abundant. In E-Indonesia only very deep at the base of outer reef drop-offs. In Japan only shallow. Male develops pointed fins and a dark patch behind eye. Below: **Lemon peel** *C. flavissimus*, to 14 cm, entire area, 1-10 m. No blue eye ring in Indian Ocean.

Centropyge heraldi **Bali, Indonesia**

Blue-and-gold angelfish
Le: To 16 cm. Di: West Pacific. De: 3-20 m. Ge: Common in rich coral and rocky coastal reefs and lagoons. Adults often in small aggregations, small juveniles secretive. Below: **Damsel pygmy angelfish** *C. flavicauda*, to 8 cm, widely distributed, throughout entire area, 10-60 m.

Centropyge bicolor **Ambon, Indonesia**

Centropyge bispinosa **Flores, Indonesia**

Coral beauty

Le: To 10 cm. Di: Entire area. De: 3-60 m. Ge: Common, deep but also in clear coral-rich lagoons. Highly variable, blue to red or very pale with thin vertical lines. In pairs or loose groups. Below: **Rusty angelfish** *C. ferrugata*, to 10 cm, Philippines to S-Japan, 6-30 m. See also p.150-151.

Centropyge tibicen **Kerama I., Japan**

Keyhole angelfish

Le: To 15 cm. Di: Entire area. De: 3-35 m. Ge: Coastal reefs, in various habitats. Small juveniles black with white lateral spot which varies greatly in size. Large individuals obtain a blue sheen. Below: **Pearly-scale pygmy angelfish** *C. vroliki,* 12 cm, entire area, 3-25 m. On rocks with algae, sponges.

Centropyge loricula **Nabucco I., Indonesia**

Flame pygmy angelfish

Le: To 10 cm. Di: Entire area. De: 3-20 m. Ge: Most attractive of all pygmy angelfishes. Rare in coastal reefs, more common at oceanic locations. Below: Similar **Shepard's pygmy angelfish** *C. shepardi,* to 10 cm, entire area, 4-22 m. Difficult to approach, stay close to their coral refuge.

Japanese pygmy angelfish

Le: To 15 cm. Di: In our area restricted to southern Japan, especially common at Izu Peninsula. De: 15-60+ m. Ge: Found on coastal rocky reefs, also in areas with moderate coral growth. Only genus member with sexual colour dimorphism. Below: Pair.

Centropyge interrupta **Ogasawara Is., Japan**

Many-banded angelfish

Le: To 12 cm.
Di: Widespread in West Pacific.
De: 3-70 m.
Ge: Secretive in caves and ledges along deep drop-offs. Sometimes in shallow reef cuttings. Often seen in small groups in large caves, upside-down on the ceilings. Small juveniles have a large ocellus posteriorly on dorsal fin. This species is distinct from the other pygmy angelfishes and is placed into its own genus.

Paracentropyge multifasciata **Bali, Indonesia**

Blue-backed angelfish

Length: To 12 cm. Distribution: Philippines north to Japan. Depth: 6-40 m. General: Prefers protected areas with clear water. Swims almost always upside-down in caves. Below: Hybrid between *Paracentropyge multifasciata* and *Paracentropyge venusta*.

Paracentropyge venusta **Okinawa, Japan**

Pygoplites diacanthus **Derawan I., Indonesia**

Regal angelfish

Le: To 25 cm. Di: Entire area. De: 3-80 m. Ge: Common in rich coral areas from shallow coastal reefs to outer reef drop-offs. Small juveniles (below) in caves along drop-offs. Regional colour variation: Indian Ocean form is yellow or orange on chest instead of grey in the West Pacific form.

Genicanthus semifasciatus **Okinawa, Japan**

Japanese swallow

Le: To 21 cm. Di: Northern Philippines and Taiwan to S-Japan. De: 15-100+ m. Ge: Outer slopes of rocky and coral reefs. All genus members with distinct sexual colour dimorphism. Below: **Black-spot angelfish** *G. melanospilos,* to 18 cm, entire area, 20-50 m. Often in small groups dominated by the male.

Genicanthus takeuchii **Ogasawara Is., Japan**

Takeuchi's angelfish

Le: To 25 cm. Di: Only known from Bonin Is. and Marcus I., NW-Pacific. De: 36+ m. Ge: Only recently described. Below: **Lamarck's angelfish** *G. lamarck,* to 22 cm, entire area, 10-50 m. Unlike most angelfishes often in midwater above reefs. Male with harem like seen in other genus members.

Damselfishes are a very large family, particularly in tropical coral reef habitats, and some species occur in such great numbers that they are probably the most numerous fishes in those areas. An estimated 300 species are globally distributed in tropical and subtropical waters. The species are placed in three subfamilies: Amphiprioninae, comprising the anemonefishes; Chrominae, species of the genera *Chromis, Acanthochromis* and *Dascyllus;* and Pomacentrinae including the remaining genera. Most genera are distinct in shape and other characters, but within a genus many species look similar and also have geographical variations. Juveniles are often very different from the adult, whilst differences between sexes are small and often colour changes occur only during the spawning period.

Habitats are nearly always reefs which feature plenty of cover such as small crevices close to main food sources. Basically, algae feeders occupy the shallow reef flats and planktivores the rocky outcrops. Only a few species move about over open substrate.

Spawning varies from spawning in pairs to community spawning in large groups. The demersal eggs vary in shape from ovoid to elliptical and in size from about 1-3.5 mm long. Hatching often occurs after a few days and only one species *(Acanthochromis polyacanthus)* has non-pelagic young. Postlarvae settle at various sizes from about 10-20 mm, depending on the species. Some juveniles have an extended pelagic stage and seek shelter under floating weeds.

Amphiprion polymnus **Palau**

Panda clownfish

Length: To 13 cm.
Distribution: West Pacific.
Depth: 3-35 m.
General: Symbiotically associated with the anemones *Heteractis crispa* and *Stichodactyla haddoni,* the latter often well away from reefs on sandy or muddy substrates including deep channels between reefs. Highly variable within local groups and geographically. Also known as Saddleback anemonefish a name that could be applied to many species.

Below: A very unusual coloration of this species, probably a hybrid with *A. melanopus.* Photo Philippines.

Spine-cheek anemonefish
Le: To 14 cm. Di: Widespread West Pacific. De: 3-15 m. Ge: Protected coastal reefs, only found in the anemone *Entacmaea quadricolor.* Male small, female usually dark but orange in some areas. Obvious spine crossing white head band. Below: Colour variation from Sumatra.

Premnas biaculeatus **Bunaken I., Indonesia**

Black anemonefish
Le: To 11 cm. Di: Eastern Indonesia, from Bali to Micronesia. De: 3-15 m. Ge: Mainly associates with *E. quadricolor* but also found in *H. crispa* and *H. magnifica.* In large numbers where anemones congregate. Below: **Tomato anemonefish** *A. ephippium*, to 12 cm, Malaysia to Java, 3-15 m.

Amphiprion melanopus **Banda Sea, Indonesia**

Bridled anemonefish
Le: To 14 cm. Di: West Indonesia, Malaysia, Thailand, Vietnam, Philippines. De: 2-18 m. Ge: Adults with a single white bar, juveniles with two or three. Similar to *A. melanopus,* but has red ventral and anal fins. Associates only with *Entacmaea quadricolor*. Below: Fish with eggs.

Amphiprion frenatus **Cebu, Philippines**

Amphiprion chrysopterus **Palau**

Orange-fin anemonefish

Length: To 14 cm.
Distribution: Philippines, north-eastern Indonesia, ranging to Micronesia and central Pacific islands.
Depth: 3-30 m.
General: Associates with a large variety of anemones: genera *Entacmaea, Heteractis* and *Stichodactyla*, and the colour of the fish varies slightly in relation to different anemones. The pectoral fin is always orange or yellow but the ventral fins vary from orange to black. Clear water reefs, often in surge zones and found mainly in pairs.

Amphiprion sandaracinos **Sangihe I., Indonesia**

Eastern skunk anemonefish

Le: To 12 cm. Di: Indonesia to Philippines. De: 6-20 m. Ge: Associates with *H. crispa* and *S. mertensii.* Adults in pairs, often with juveniles sharing anemone. Coastal protected reefs. Below: **Pink anemonefish** *A. perideraion,* to 18 cm, entire area, 3-15 m.

Amphiprion clarkii **Alor I., Indonesia**

Clark's anemonefish

Le: To 14 cm. Di: Thailand to Philippines. De: 1-20 m. Ge: Usually black with variable amount of orange on head, ventral parts and fins. Three white bars on head, body and base of caudal fin. Host anemone often *S. haddoni.* Much less host-specific (with 10 spp.!) than other anemonefishes.

Western clown-anemonefish

Length: To 8 cm.
Distribution: Widespread in tropical West Pacific.
Depth: 1-15 m.
General: A common species in our area, associating with *Heteractis magnifica* (see photo right, anemone has retracted, associated animals are clearly visible), *Stichodactyla gigantea* and *S. mertensii.* Coastal protected reefs and usually in small groups.

Amphiprion ocellaris **Ambon, Indonesia**

Eastern clown-anemonefish

Length: To 8 cm.
Distribution: West Pacific, Vanuatu, northern New Guinea to Sulawesi, Indonesia.
Depth: 1-10 m.
General: Associating with *Heteractis magnifica, Stichodactyla gigantea* and *S. mertensii.* Coastal protected reefs and usually shallow, in small groups. Highly variable, especially when in *Stichodactyla* anemones, with black coloration.

Amphiprion percula **Sulawesi**

Dascyllus melanurus **Flores, Indonesia**

Black-tail damsel

Le: To 8 cm. Di: West Pacific, Indonesia to Philippines. De: 1-10 m. Ge: On protected coastal reefs and in lagoons. Small groups with *Acropora* corals. Below: The similar **Humbug damsel** *D. aruanus* has no black tail. It is common throughout our area.

Dascyllus reticulatus **Bali, Indonesia**

Reticulated damsel

Le: To 6 cm. Di: Tropical West Pacific. De: 3-30 m. Ge: In small to large groups with *Acropora* heads on rock or near dead corals. Below: **Three-spot damsel** *Dascyllus trimaculatus,* to 8 cm, entire area, 1-50 m. Juveniles often in great numbers with anemones, sometimes sharing it with anemonefishes.

Amblyglyphidodon aureus **Bunaken I., Indonesia**

Golden sergeant

Le: To 12 cm. Di: Entire area. De: 6-30 m. Ge: Coastal to outer reefs, eggs are laid on seawhips and defended by male. All yellow, body deep. Little variation with growth or between sexes. Below: **Black-and-yellow damsel** *Neoglyphidodon nigroris,* to 12 cm, tropical West Pacific, 3-30 m.

Spiny-tail puller
Le: To 14 cm. Di: Indonesia to Phillipines. De: 2-40 m. Ge: Clear shallow coastal lagoons to deep outer reef drop-offs. Extremely variable between different areas from totally dusky to pale green or posteriorly white. Unique among damsels, closely guards young. Below: Colour variant.

Acanthochromis polyacantha **Flores, Indonesia**

Yellowtail blue damsel
Le: To 5 cm. Di: West Pacific, patchy. De: 1-15 m. Ge: Quiet coastal bays, in hard corals, numerous on reefs. Below: **Bleeker's damsel** *C. bleekeri*, to 6.5 cm, Indonesia to Philippines, 6-30 m. Steep current-swept coastal rubble slopes. Known from few localities, may prefer deep habitats elsewhere.

Chrysiptera parasema **Talaud I., Indonesia**

Blue-green chromis
Le: To 8.5 cm. Di: Tropical West Pacific. De: 1-20 m. Ge: In lagoons and coastal reefs in small schools, feeding closely above the ground on zooplankton. Below: Similar **Green chromis** *C. viridis* is distinct by blue-green colour and, from *C. atripectoralis*, by lack of dark spot near pectoral fin base.

Chromis atripectoralis **Sipadan I., Malaysia**

Chromis lepidolepis **Palau**

Scaly chromis

Length: To 8 cm.
Distribution: Entire area.
Depth: 1-14 m.
General: This little damselfish lives in coral reef areas, on coastal as well as on outer reefs and slopes. The black margin on its dorsal fin, the fine white tips of each dorsal fin spine, and the black margins of the tail fin are distinct. The species feeds on zooplankton.

Below: **Vanderbilt's chromis** *C. vanderbilti,* to 6 cm, patchy distribution in West Pacific, 2-20 m. Commonly found in areas which are exposed to currents and swell where sufficient zooplankton is carried for the fishes to feed on.

Abudefduf vaigensis **Banda Sea, Indonesia**

Sergeant major

Le: 17 cm. Di: Entire area. De: 1-12 m. Ge: Occurs in groups in a variety of habitats ranging from rocky lagoon shorelines to reef flats, lagoon reef margins, and seaward reefs. Juveniles often in tidepools. Feeds on zooplankton, benthic algae, invertebrates. Photo left shows it together with many other damsels. Bottom: **Scissor-tail sergeant** *A. sexfasciatus,* to 14 cm, entire area, 1-15 m.

Yellow-tail sergeant
Le: To 17 cm. Di: Entire area. De: 1-30 m. Ge: Shallow coastal habitats from high-energy to quiet areas near freshwater run-offs, also in current channels, schools in deep water. Changes to pale grey when swimming off the ground. Below: **Black-tail sergeant** *A. lorenzi,* to 25 cm, Sulawesi, Palau, PNG, 10-60 m.

Abudefduf notatus **Mindoro, Philippines**

Philippine damsel
Le: To 10 cm. Di: Entire area. De: 1-15 m. Ge: Highly variable with different forms in different areas. Coastal to outer reef crests on walls of shallow drop-offs in cuttings and channels, usually singly. Below: **Princess damsel** *P. vaiuli,* to 8.5 cm, entire area, 3-25 m. In various habitats, highly variable.

Pomacentrus philippinus **Hong Kong**

Dick's damsel
Le: To 10 cm. Di: Tropical West Pacific. De: 1-15 m. Ge: Clear coastal reefs, lagoons, to outer reefs in coral heads. In various sizes in small groups. Little variation with growth or area. Below: **Jewel damsel** *P. lacrymatus,* to 10 cm, tropical West Pacific, 3-15 m. Very territorial and aggressive.

Plectroglyphidodon dickii **Derawan I., Indonesia**

ANEMONE-CITY

I have been looking forward to this dive site in eager anticipation, because those few underwater photographers who have been there already all agreed: You will see the largest field of sea anemones in the oceans anywhere, a field so large that it is measured not in square metres but in square kilometres; you will find sea anemones not only in shallow water but also down to a depth of 40 m. But be cautious, because currents there can set in without any warning! Yet, even such explicit details did not prepare me for the magnitude of all this; after all, by comparison the anemone-covered panorama reef in the Red Sea would only be a dandelion in a meadow! On board the 'BARUNA ADVENTURER' we approached the island of Pura (east of the Indonesian Nusa Tengara Islands) already the night before. The sea surface was as smooth as millpond and it was dead calm. I had intended to photograph there for at least an entire day.

PHOTOS: HELMUT DEBELIUS

As far as one can see: The field of sea anemones off the Indonesian island of Pura (Nusa Tengara, Lesser Sunda Isl.), depth 8 m.

First thing in the morning at 7 o'clock - off the vast beach of Pura Island - we dove in to the shallow water. I looked down and was surprised to note that there were hardly any corals; below the surface the rocky beach of the island sloped gently downward. The rocks were no longer visible because all of them - as far as the eye could see in good underwater visibility - were covered by sea anemones. And, of course, there were the inevitable clownfishes, living in their respective host anemones. An unbelievable sight! How could I convey the magnitude of this field of anemones in photographs? At that time the sun had not yet risen above the island behind us and I needed bright sun light urgently for some wide-angle exposures. Occasionally, a colourful sponge, a curled up gorgonian head or an unknown algae species would protrude above the tentacle forest, largely made up of Knobby Sea Anemones, *Entacmaea quadricolor.* A slight lateral current did not interfere with our photographing activities and so we dove down to a depth of 30 m, where the slope started to drop off steeply.

There too the typical anemone carpet continued on in the - by then - rather cool but clear water. Here the anemones even covered steep-walled rocks. All around us where school of banner fishes and surgeon fishes, and groups of yellow damsel fishes would come darting out of rocky crevasses. Here, at the 30 m zone, there was far more marine life than in the shallow water above! Artificial light was insufficient to captured this imposing sight on film: The sun had still not yet risen above the mountains of the island. Swimming up the slope we took our time to have a close look at the sea anemones and their clownfishes. Apart from Knobbies, I saw a few red coloured specimens of the Carpet Sea Anemone, *Stichodactyla haddoni,* and the colourful Beaded Sea Anemone, *Heteractis aurora.* I quickly realised that all the clownfishes were of a single species only, Clark's Clownfish, *Amphiprion clarkii.* Pairs of these fishes occupied many of the anemones, but there were far more sea anemones than clownfishes, to occupy them all. By that time the dive had already lasted nearly an hour, when - while decompressing at a depth of 6 m - a Yellow-lipped Sea Krait, *Laticauda colubrina,* slithered among the hydroids over the carpet of anemones, without displaying any apparent stinging effect. What an extraordinary dive it had been!

The second dive is planned for 10 o'clock, and the boat was moved about a kilometre further off-shore from Pura Island. According to the dive-master of the 'BARUNA ADVENTURER', this was to a site were there are supposed to be even more sea anemones. Unbelievable! At that time the sun was up and I was excited and ready to go. The films had been changed in both cameras, and I was waiting for the signal to permit me to jump off the vessel's diving platform in to the water. Waiting next to me were other 'anemone lovers'. But what was that suddenly appearing at the surface? Strong, river-rapid-like tidal currents and whirlpools, which had already given us a severe scare while diving off Alor Island, emerged.

Immediately, the dive-master changed his plan, and with the aid of an inflatable dingy and while snorkelling we tried to find a calmer dive site off the coast of Pura Island; yet, anywhere along 'sea anemone city' a strong current had set in. Only in a bay near a fishing village diving seemed feasible: Photographers in to the water first! There the slope drops off more rapidly and it is covered with fish traps made of bamboo. It is was also getting rather dark by then because the sun has disappeared behind some black clouds.

Even in a depth of 30 m the slope off Pura is largely covered with sea anemones.

At a depth of 15 m, and with my legs firmly pressed against a rock, I am starting to photograph the sea of tentacles stirred up by the current. The slave flash unit, held by my diving partner, is to enhance the ambient lighting situation. I can still see it as a large fish trap is being raised up vertically in the water, when a sudden lateral current sets in. With the dome port pressed firmly against my body, I manage to grab a secure hold with one hand. I must not drift downwards! It is impossible to move otherwise my mask could be pulled away from my face. We cling tightly to the side of a large rock, where we are less buffeted by the whirlpool current. A few minutes later the spook is over and with some relief we continuing photographing. The sea anemones that had been churned up by the current have settled down again to their rocky substrate. We are pleased to be able to recommence our activities, but the apparent tranquility was misleading. Suddenly the 'tsunami' started up again from different directions and once again we have to seek cover behind a large rock. There is no way we can continue photographing; in fact, there is a real danger now that the cameras might be smashed against the rock if we are not careful. After a few minutes we ascend - dispirited - back through open water, and are quickly picked up by the support dingy. It was supposed to have been a 'day of anemones', but Baruna, the God of the Sea, apparently had different ideas. Nevertheless, all participants of this diving expedition will retain fond memories of the city of sea anemones. Nobody has ever seen such a vast carpet of sea anemones before!

A Yellow-lipped Sea Krait *Laticauda colubrina* slithers among the hydroids.

The vast anemone carpet is made up of Knobby Sea Anemones *Entacmaea quadricolor*.

The hawkfishes are a tropical fish family, 9 genera and 35 species are known, which are mostly distributed in the Indo-Pacific, only 3 species occur in Atlantic waters. These small fishes hug the bottom, perched on their thickened lower pectoral fin rays, but unlike most bottom dwellers they are very active, often restlessly changing position. Only one species regularly swims above the substrate to feed on zooplankton. Most species live in the shallows, on reef crests and in areas with a strong surge, some can only be found deeper than 30 m. Most species are habitat-specific, they are found living with certain sponges or corals and usually occur in loose aggregations. They are carnivores, feeding on small fishes and invertebrates. Hawkfishes are generally small and less than 10 cm long but a few attain almost 30 cm in length.

Hawkfishes are pelagic spawners, their postlarvae are fairly large, usually in excess of 20 mm and up to about 40 mm long. Filaments at the tip of the dorsal fin spines, often tufted, are diagnostic for the family.

Cirrhitichthys aureus **Kerama I., Japan**

Golden hawkfish

Le: To 15 cm. Di: Entire area. De: 20-60 m. Ge: Colour yellow to brown, sometimes with dark blotches similar to *C. aprinus* (below, to 12 cm, entire area, 5-40 m). On sponges along outer reef drop-offs, usually in depths of 20+ m and not common anywhere, only observed singly.

Cirrhitops fasciatus **Singapore**

Redbanded hawkfish

Length: To 11 cm.
Distibution: Entire area.
Depth: 1-30 m
General: Occurs in lagoons or seaward reefs with clear water. Prefers to stay at the base of coral heads, not on top.

Coral hawkfish
Le: To 10 cm. Di: Entire area. De: 5-40 m. Ge: Clear coastal reef crests and rubble slopes to deep outer reef habitats. Juveniles singly, adults usually in pairs. Colour slightly varies between habitat with pale specimens from white sand reefs to darker colours on darker reefs. Below: Colour variant.

Cirrhitichthys falco **Cebu, Philippines**

Spotted hawkfish
Le: To 10 cm. Di: Widespread tropical Indo-Pacific, entire area. De: 3-40 m. Ge: Shallow coastal rocky reefs to deep rubble slopes. Usually observed singly. Colouration varies from dark to light to suit various habitats, sometimes bright red in deep habitats. Below: Colour variant.

Cirrhitichthys oxycephalus **Hong Kong**

Flame hawkfish
Le: To 9 cm. Di: Indonesia to Philippines. De: 5-37 m. Ge: Coloration distinct. Below: **Lyre-tail hawkfish** *Cyprinocirrhites polyactis,* to 10 cm, entire area, 15-50 m. Coastal to outer reef slopes, often in groups at remote outcrops, in current-prone areas. Feeds on plankton well above substrate.

Neocirrhites armatus **Luzon, Philippines**

Paracirrhites arcatus **Bunaken I., Indonesia**

Ring-eyed hawkfish
Le: To 14 cm. Di: Entire area. De: 1-10 m. Ge: Mainly on shallow outer reef crests, perched on outer coral branches. Often in small groups per coral, numerous individuals in neighbouring corals. Dark brown to grey and red, white band on side sometimes indistinct. Below: Colour variant.

Paracirrhites forsteri **Ambon, Indonesia**

Forster's hawkfish
Le: To 20 cm. Di: Entire area. De: 3-30 m. Ge: Clear coastal to outer reef slopes, usually perched among the outer-most coral branches to observe surroundings. Sometimes all dark with yellowish tail and adults with additional spotting on head. Below: Colour variant.

Oxycirrhites typus **Tukang Besi, Indonesia**

Longnose hawkfish

Le: To 10 cm. Di: Widespread in Indo-Pacific, entire area. De: 5-70 m. Ge: Primarily a deep water dweller but in some areas as shallow as 5 m, where black coral or gorgonians grow which they seem to associate with.
Below: Colour variant.

JAWFISHES — OPISTOGNATHIDAE

Randall's jawfish
Le: To 11 cm. Di: Indonesia to Philippines. De: 3-20 m. Ge: Common from Bali to Flores. See also p.43-44.

Jawfishes are a family of small burrowing species, many of which are undescribed, comprising 3 genera and an estimated 70 species. Distributed in all tropical waters except eastern Pacific. Somewhat odd-looking fishes with large eyes, placed high and forward on head, a large mouth used for incubating eggs, and elongate bodies. The burrows are usually made (photo right) vertically in sand, and are reinforced with small rocks or coral bits. Usually these fishes are found with the head just peeking out of the burrow, looking out for passing plankton, and retreating into the burrow tail-first. Below: Two yet undescribed *Opistognathus* spp.

Opistognathus randalli **Mabul I., Malaysia**

BANDFISHES — CEPOLIDAE

Bandfish
Le: To 20 cm. Di: Indonesia. De: 6-40 m. Ge: Common in Flores. In small to large groups on mud or sand flats and slopes, coming out to feed when currents carry plankton.

The Bandfishes are a little-known family of small fishes with at least 4 genera and about 20 species widely distributed in all oceans. Most of these eel-shaped fishes are coloured red or pink. They primarily inhabit mud or fine sand substrates in self-made burrows, in the range of 20-500 m and feed on plankton.

Cepola **sp.** **Flores, Indonesia**

PORTABLE BURROWS AND THE SECOND HAND HOME MARKET

In the third century B.C., Aristotle, stroking his long white beard, probably pondered the mysteries of the hermit crab while sitting on the shores of the Mediterranean. "The hermit crab grows spontaneously out of soil and slime,....", he stated in his *Historia Animalium*. No doubt he wondered how this odd lobster-like crab fashioned its residence out of a marine snail's shell; and where had the original inhabitant, the slimy snail, gone to? His early observation on the origins of the hermit crab may have made sense to him, but since then many minds and untold hours of observation have been spent on the biology of these intriguing animals. One of those observers is Clay Bryce.

PHOTOS: CLAY BRYCE

Once the home is selected then it must be made to express the occupant's individualism! In this case, anemones specially selected and strategically placed by the hermit.

Hermit crabs belong to the phylum Crustacea along with crabs, succulent lobsters and feather-footed barnacles. In the past, Crustacea was considered a class or superclass of the phylum Arthropoda, which also included insects, spiders and centipedes. Their independent status is a controversial taxonomic change not recognised by all researchers. However, the Crustacea as a group, with over 30,000 species worldwide can be readily identified by their unique two sets of antennae, a hard external skeleton and segmented legs. The bodies of crustaceans are divided into two segments - the front or "cephalothorax", and the back, appropriately called the abdomen. Although at times it can be difficult to discern the latter!

It's the hermit crab's abdomen that makes it so distinct - soft, fragile and protected by a hand-me-down gastropod shell. It is relatively large and asymmetrically coiled, and has a modified tail fan used to grasp the shell. The left wing of the tail fan is larger than the right and both have roughened surfaces, which when pressed against the inside of the shell give considerable grip. Along the left side of the female's abdomen are the pelepods, small appendages with fine hairs used to hold eggs.

Like all crustaceans, hermits moult their calcareous "skin" as they grow. Once the old cuticle is discarded, leaving a new soft layer, the hermit retreats inside the shell to await the hardening process. In some species the claw is moulded to perfectly fit the shell entrance. The "claw door" acts to seal off the shell against predators or, in the case of land hermits, the effects of desiccation. This is achieved during the first moult in a new shell when the crab pulls the soft claw firmly against the opening. As the claw hardens during calcification it is moulded into a snug fit. This process of moulting can only continue for so long, soon the animal has outgrown it's quarters and must look for a new home.

To choose a new shell is no simple matter - just like buying a new house, you must consider the look, the feel and the size before deciding its suitability. A hermit crab will investigate a

Nakedness could mean death from a hungry predator as this hermit crab scuttles to find a new home.

number of possibilities as it moves about in its quest for a new home. Potential shells are closely inspected. It is first touched with a claw or leg to check if anyone is home. If there is no response an antennae is used to carefully probe inside. An un-occupied shell then undergoes a rigorous pre-selection process. The hermit will determine suitability of the shell surface, whether the shell is easily moved, what the internal size is and if weight and aperture dimensions are right. In fact, they can determine the difference in weight between two shells of the same species, ensuring they end up with a home that's not too heavy.

A stout shell makes for a safe home, while a colourful occupant makes a fashion statement.

At the completion of this selection process the hermit will scuttle alongside the empty shell, quickly withdraw its vulnerable abdomen from its used mobile home and carefully insert it into the new. If the fit is not quite right the crab will continue to investigate other possibilities. The selection of the right shell is an instinctive trait fully developed in juveniles rather than learned through trial and error. Competition for available shells can be fierce with injuries including lost claws, legs and antennae. A lost appendage, although replaced during the next moult, could have serious consequences for a hermit crab especially if it's a claw used as a front door!

Without a shell the hermit reveals an essential body part - the left wing of the tail fan is larger than the right and both have roughened surfaces that when pressed against the inside of the shell give considerable grip. Along the left side of the female's (right) abdomen are the pelepods, small appendages with fine hairs used to hold eggs.

A new home for a hermit caught out in the cold. It will first investigate the shell with an antenna to see if anybody is home.

The wrasses make up one of the largest families of fishes on tropical reefs, worldwide it comprises more than 60 genera and an estimated 400 species. The various genera can be divided into several distinct groups but their exact relationship is often not clear. Some genera too have species groups which may represent subgenera or another valid genus. Much research is needed to clarify the situation. About 150+ species occur in our area, many of which only recently described and some are yet to be named. This number will probably increase as exploration of new areas and investigations into the family continue.

Wrasses are highly diverse in shape and vary greatly in size. Nearly all wrasses have various and distinct colour forms during growth, from the postlarval stage to the adult, and between sexes. Normally juveniles become female first and males derive from females. In almost all species the fully developed males become very colourful and can look totally different from the earlier stages. Most males dominate a number of females and are territorial towards other males. If a male disappears from a group, the next most dominant adult, usually largest and normally still female, takes over and changes sex (protogynous hermaphrodites). They are mostly benthic feeders, taking a great variety of invertebrates but some prefer small fishes or eggs. A few are specialised feeders, taking plankton or picking parasites from other fishes. All species are diurnal and retire by either burying in the sand or by wedging themselves into crevices. Wrasses are primarily reef dwellers but a few species have adapted to open sand or rubble flats and may occur deep. All are thought to produce pelagic eggs. Postlarvae settle at various sizes, depending on species, from about 6-15 mm.

Bodianus anthioides **Bali, Indonesia**

Lyretail hogfish

Le: To 22 cm. Di: Entire area. De: 6-60 m. Ge: Coastal to outer reefs in rich coral growth along drop-offs. Usually in depths in excess of 20 m but if suitable habitats are found much shallower also seen in 6 m. Often with large gorgonian fans. Juveniles (below) may clean parasites off other fishes.

Bodianus mesothorax **Bohol I., Philippines**

Blackbelt hogfish

Le: To 20 cm. Di: Widespread in tropical West Pacific. De: 5-40 m. Ge: Mostly found on clear outer reefs in rich coral growth but in some areas they are just as common on coastal reefs. Juveniles dark brown with large yellow spots, secretive in caves. Juvenile below.

Diana's hogfish

Le: To 24 cm. Di: Entire area. De: 3-40 m. Ge: West to Java, retains spots on fins (looses spots in Indian Oc.). Outer reefs in rich soft coral growth, from shallow reef flats to deep drop-offs. Juveniles in 20-30 m, upside down under overhangs. Below: **Yellow hogfish** *B. bimaculatus,* to 10 cm, entire area, 20-60 m.

Bodianus diana **Hong Kong**

Eclipse hogfish

Le: To 40 cm. Di: NE-Indonesia to S-Japan. De: 3-40 m. Ge: On coastal reefs, also outer reefs of oceanic localities. Juveniles similar to adult but lines formed by small spots and larger posterior black area on body. Below: **Saddleback hogfish** *B. bilunulatus,* to 55 cm, entire area, 10-100 m.

Bodianus loxozonus **Flores, Indonesia**

Cigar wrasse

Le: To 48 cm. Di: Entire area. De: 1-30 m. Ge: Coloration varies little between stages. In seagrass beds but also on reefs with soft corals where they are often bright yellow. Usually shallow but also on slopes to 30 m. Below: **Yellow wrasse** *H. chrysus,* to 12 cm, West Pacific to Java, Indonesia, 3-30 m.

Cheilio inermis **Sumatra**

Halichoeres marginatus **Mabul I., Malaysia**

Dusky wrasse
Le: To 16 cm. Di: Entire area. De: 3-30 m. Ge: Coastal reefs from shallow reef flats and slopes. Small juveniles in crevices, often with sea urchins. Below: The common **Hoeven's wrasse** *H. melanurus*, to 12 cm, entire area, 1-15+ m. Patch reefs in lagoons, and coral areas near sand.

Halichoeres richmondi **Kerama I., Japan**

Richmond's wrasse
Le: To 14 cm. Di: Tropical West Pacific. De: 3-20 m. Ge: Very similar to the following species. Below: **Chain-lined wrasse** *H. leucurus*, to 14 cm, Indonesia to S-Japan and New Guinea, 3-15 m. Long snout distinctive. In soft coral gardens on reef flats, upper edge of drop-offs.

Halichoeres melasmapomus **Sumatra**

Blackear wrasse
Le: To 24 cm. Di: Widespread in tropical West Pacific, Indonesia to Philippines. De: 10-55 m. Ge: On sand and rubble on outer reef slopes. Often below 30 m at the base of steep slopes. Below: **Timor wrasse** *H. timorensis*, to 14 cm, Malaysia, Indonesia, PNG, 1-7+ m. On shallow coastal reefs.

Leopard wrasse
Le: To 14 cm. Di: Indonesia to S-Japan. De: 4-30 m. Ge: In lagoons and on outer reefs on mixed ground. Below: **False leopard wrasse** *M. ornatus,* to 13 cm, Indonesia, 3-25 m. Replaced by similar *M. meleagris* (above) north of Manado and in easternmost parts of Indonesia. Coastal reefs, outer reef crests.

Macropharyngodon meleagris **Mindanao, Philippines**

Chiseltooth wrasse
Le: To 25 cm. Di: Entire area. De: 3-40 m. Ge: Occurs singly primarily in coral-rich areas of outer lagoon channels and seaward reefs. Juveniles (below) seem to prefer somewhat deeper water, often seen at drop-offs below about 20 m. Feeds on hard-shelled bottom-dwelling invertebrates.

Pseudodax moluccanus **Palau**

Blue-ribbon wrasse
Le: To 14 cm. Di: West Pacific. De: 3-20 m. Ge: Mainly along outer reefs in small to large aggregations of mixed sex. Brightly coloured males more obvious than females, often swimming high above substrate. Below: The common **Silver-streaked wrasse** *S. strigiventer,* to 14 cm, entire area, 1-6 m.

Stethojulis trilineata **Mabul I., Malaysia**

Thalassoma lutescens **Palawan, Philippines**

Yellow moon wrasse
Le: To 25 cm. Di: West Pacific. De: 3-30 m. Ge: Common in cooler part of range. Outer reefs in lagoons and slopes to 30 m. Small juveniles singly but usually in small groups when becoming older. Adults in some areas numerous. Below: **Moon wrasse** *T. lunare,* to 25 cm, entire area, 1-25 m.

Oxycheilinus bimaculatus **Banda Sea, Indonesia**

Two-spot splendour wrasse
Le: To 15 cm. Di: Tropical West Pacific. De: 2-110 m. Ge: On rubble, in seagrass. Left: Two males contend for a female and a spawning place. Below: **Napoleonfish** *Cheilinus undulatus,* to 230 cm, entire area, 1-60 m. See also MOURNING FOR NAPOLEON on p.54-57.

Epibulus insidiator **Derawan I., Indonesia**

Slingjaw wrasse

Le: To 35 cm. Di: Widespread in tropical Indo-Pacific. De: 3-40 m. Ge: Inhabits coastal to outer coral reefs, usually seen singly. Adults are often shy. Presently a single species is recognised but the genus may comprise three species. Left: Male. Below: Female.

Dotted wrasse

Le: To 13 cm. Di: Entire area. De: 2-32 m. Ge: Males have very long pelvic fins, both sexes have bodies covered with fine pink or blue dots. Males territorial, with harems. Usually seen just off the bottom. Eats plankton. Below: **Blue-head fairy wrasse** *C. cyanopleura*, 13 cm, Indonesia to S-Japan, 1-35 m.

Cirrhilabrus punctatus **Banda Sea, Indonesia**

Red-head fairy wrasse

Le: To 13 cm. Di: Indonesia. De: 3-20 m. Ge: On coastal reefs, usually in small to large groups, colourful males greatly outnumbered by other phases. Below: **Lubbock's fairy w.** *C. lubbocki,* to 8 cm, Celebes, Philippines, 4-45 m. Seen in groups above finely branching coral and over rubble bottoms.

Cirrhilabrus solorensis **Bali, Indonesia**

Filamented flasher wrasse

Le: To 10 cm. Di: Indonesia, Philippines. De: 6-35 m. Ge: Extremely variable, males change colour during display. Reef slopes in coral-rich areas with rubble, usually in great numbers feeding high above substrate on plankton. Below: *Paracheilinus* sp., also from Flores.

Paracheilinus filamentosus **Flores, Indonesia**

Choerodon fasciatus **Okinawa, Japan**

Harlequin tuskfish
Le: To 25 cm. Di: Taiwan to Ryukyus, very rare in PNG. De: 1-15 m. Ge: Usually close to shallow coral walls in lagoons. Sought after by photographers and for aquaria because of stunning colours. Juveniles in crevices. Below: **Yellow-cheek tuskfish** *C. anchorago*, to 38 cm, 2-25 m.

Novaculichthys taeniourus **Batangas, Philippines**

Rock-mover wrasse
Le: To 30 cm. Di: Tropic. W-Pacific. De: 3-20 m. Ge: Adults are experts in turning over large pieces of dead coral or rocks for a feed, working in pairs with one lifting, the other taking care of prey, and taking turns. Below: The well-known **Cleaner wrasse** *Labroides dimidiatus,* to 8 cm, entire area, 1-50 m.

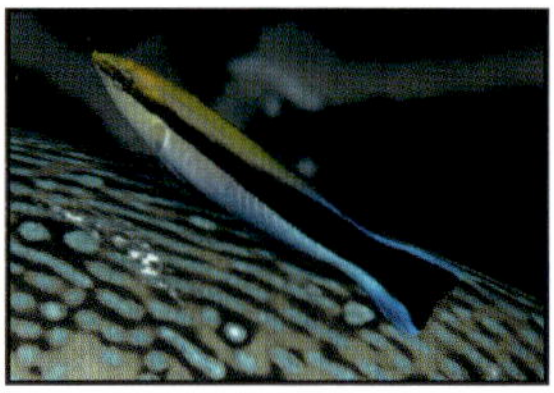

Xyrichthys pentadactylus **Sulawesi**

Red-spots razorfish
Le: To 20 cm. Di: Entire area. De: 1-20 m. Ge: Small juveniles uniformly brown, white, bright yellow or green to almost black. Common on coastal mud and sand slopes from intertidal downward. Below: **White-blotch razorfish** *X. aneitensis,* to 20 cm, tropical West Pacific, 1-25 m.

LAST OF THE REEF BUFFALO

If the scenario is reminiscent of the last days of the great herds of bison on the American plains, the analogy is even more appropriate for another fish – the bumphead parrotfish, *Bolbometopon muricatum*. The comparison to a herd of buffalo munching across the prairie is inescapable to anyone who has observed this spectacle of a school of bumpheads grazing over a reef, right down to the noise and clouds of dust. The noise comes from the demolition of coral – both living and dead – as it is bitten off by teeth fused into parrot-like beaks, and then crushed by the impressive mills of pharyngeal teeth in the throat. They have been rumoured to smash coral for feeding by ramming it with the impressive forehead hump, which becomes red in large individuals, but researchers insist that this is not the case. The function of the hump, if any, remains unknown. Underwater photographer Doug Perrine tells us more about the 'reef buffaloes.'

PHOTOS: DOUG PERRINE

To approach such a group of humphead parrotfish is only possible in the protected coral reefs of a nature reserve like Sipadan.

Dr. David Bellwood of James Cook University estimates that each fish consumes between 4-5 metric tons of reef per year, half of which is live corals – enough to have a significant effect on coral community structure. By eating the fast-growing *Acropora* corals, the parrotfish may allow the slower-growing corals more room on the reef, according to Dr. Nick Dulvy of the University of Newcastle-upon-Tyne. "Dust" clouds the water as the coral is excreted out the other end of the digestive tract. It comes out as pure white sand, after the extraction of nutrients from algae and invertebrates. Eventually piles of this fish poop wash up on shore for tourists to lounge on.

As with the Napoleon wrasse, bumpheads are often described in published references as "wary." Until recently I would have considered this term an understatement with regards to both species. The early part of my diving career was spent in islands where both were highly valued as food fish. In Ponape, the humphead wrasse was royal food – to be presented to the Namwarki, or king, in the rare event that one was captured. By day both species were only seen as phantoms at the very edge of visibility, so the only way to capture one was at night. The wrasses sleep well tucked into caves in the coral. If found, they were easily speared, but only extracted with great difficulty (in the pre-cyanide days, of course). The parrotfish present a different problem for the fisherman. Although smaller than the wrasses, attaining a size of 1.3 m and 46 kg to perhaps as much as 61 kg, they are more heavily armoured. The massive scales will deflect a spear even fired at point-blank range, bending the shaft without penetration. With little fear of predation, the parrotfish often sleep right out in the open, on patches of sand or coral ledges. Native divers learned to capture them by cupping one hand over the forehead bump, cradling the body with the other arm, and just swimming them up to the surface and throwing them into the canoe. If spearing one was hard, the thought of getting a good underwater photo of such a fish seemed an impossibility.

These parrotfish are said to smash coral by ramming it with their forehead hump but researchers reject this idea. Photo Layang Layang.

That was before I saw the great herds of Layang Layang and Sipadan in Malaysia. Completely protected from any form of fishing, and exposed to dozens or more non-lethal divers each day, these fish have developed an equanimity – a zen-like acceptance of clumsy camera-toting, flash-firing bipeds – that has to be seen to be believed. This is not to say that they can't be scared away. I've frequently seen them calmly but effectively remove themselves from the presence of agitated divers blasting them with clouds of exhaust bubbles. I remember vividly an explanation given to me by a divemaster on Sipadan as to why we hadn't seen the bumpheads at the usual place on our morning dive. He had seen another group of divers arrive at the bumphead spot before us. "They charged the parrotfish like hungry tigers!" he said, animatedly pantomiming the crazed photo divers. Bumpheads will also abandon their sleeping spots if a torch is kept on them for too long at night.

In the absence of tiger-like charges from ravenous divers, the bumphead parrotfish of Sipadan have a fairly predictable routine. In the evening the school disperses, and each fish selects a nook to sleep in on the coral drop-off in front of Borneo Divers resort. In the predawn, the fish shake the sleep out of their eyes and move a little ways down the reef lip to school up at a particular spot between the drop-off and Barracuda Point. They take turns hovering next to a barrel sponge inhabited by small cleaner fish, which come out and groom them. Shortly after sunrise the school moves up onto the reef top and spends the rest of the day circling the reef in a clockwise direction converting coral into sand. At low tide, the guides can spot them from the boat by their dorsal fins breaking the surface of the water.

The bumpheads of Layang Layang have a similar routine, but are less predictable. Occasionally we see individuals swimming around in the blue water outside the drop off, at depths down to 30 m or more. What are these fish doing by themselves out where there is no coral to eat, and plenty of sharks? The only answer I could think of is that it might have to do with reproduction. Many reef fish, like the Napoleons, do their spawning at places where currents will sweep the eggs away from the reef, and all the hungry mouths that live on it. Only one observation of spawning by bumphead parrotfish has been recorded, and it was of a single spawn by a pair of fish that rose out of the school. The fish we see alone in the blue might be ones that have just spawned and are finding their way back to the reef and the school.

But what about the bumpheads of Sipadan? After they leave their cleaning spot, they spend the day in shallow water. They can be observed at any time of day or night, yet nobody has ever seen them spawn. The published observation reports that spawning occurred near the reef edge about an hour after sunrise. It may be that the bumpheads aggregate at that particular place not just to be cleaned, but to spawn. It may also be that the arrival of multiple groups of divers each morning shortly after sunrise interferes with that ritual. Nobody knows if this species spawns once a day, once a month, or once a year, although once a month is suspected. If the giant parrotfish of Sipadan do spawn, Oakley believes that the eggs and larvae will not remain around the tiny pinnacle tip that is Sipadan, but will be swept off to a place where they have no chance of surviving to adulthood. Biological parameters for this species are mostly unknown, but "best guess" estimates are that they live up to 30 years, with sexual maturity not arriving until 6-10 years, at a size of 55-78 cm. They are usually caught at a smaller size, and therefore don't reproduce. Nick Dulvy believes that the species cannot sustain anything more than medium pressure subsistence fishing. He feels that any level of commercial fishing will drive a population to extinction, as has already occurred in Guam, parts of Fiji, and other areas. I once heard a complaint from a divemaster in the Philippines that there had been only 3 bumpheads left on his island, and the fishermen had just killed one of them. Since this is one of the few types of parrotfish where males and females do not differ in appearance, it was impossible for him to know the sex ratio of his remaining population of two fish. There is some hope

PHOTOS: HELMUT DEBELIUS

In unprotected areas like in the Philippines it is difficult for divers to approach these parrotfish.

in the observation that the Sipadan school contains some small fish. There is replenishment coming in from somewhere at some level.

Recent scientific studies have shown that marine reserves are among the most effective methods of preserving biodiversity in marine environments, and also of managing stocks of commercial fish. Some scientists now believe they are the **only** effective method of managing fish stocks, at least in the tropics. Without the reserves, there is no reproduction, leading to a complete crash of populations. Even though fishing is banned within the reserves, maintaining them creates a spillover effect from larvae drifting outside, raising fishermen's catches outside the reserves to much higher levels than before protection. Some fishermen have reported that their catches doubled in ten years after a section of their former fishing area was closed to them. Oakley's concern is that a reserve in isolation may not do the trick. It may take a series of reserves to protect the resources within an area.

The reefs of Layang Layang must be among the best-protected in the world. They flourish in splendid isolation miles from any serious sources of pollution or populations of fisherfolk. The entire atoll harbours only a dive resort, with a strict hands-off policy, and a Malaysian Navy base. The navy boys do a bit of fishing to supplement their own rations, but they make sure that no fishing boat comes anywhere close to the atoll. No cyanide or bleach touches these pristine coral heads. Sipadan's reefs are nearly as secure. And yet, fishermen in the Philippines and elsewhere may be slowly killing off the great herds of bumphead parrotfish that thrills visitors to these reefs. Once again the ocean reminds us that all things are connected, and as we destroy each part of it we are surely killing the rest as well.

At night, the humpback parrotfish rest solitarily among the corals and are easily photographed like here in Palau.

Scarus rubroviolaceus **Nabucco I., Indonesia**

Scarus rubroviolaceus **Mindoro, Philippines**

Black-veined parrotfish
Le: To 55 cm. Di: Indo-Pacific, entire area. De: 3-40 m. Ge: Juvenile with two somewhat indistinct dark spots on outer posterior scales on caudal fin and pale area heading dorsal fin. Juveniles singly on shallow to deep reef crests. Adults in pairs, usually seen mixed with other parrotfish species to feed in schools. Top left: Male. Centre: Female. See also LAST OF THE REEF BUFFALO on p.183-185 for the **Humphead parrotfish** *Bolbometopon muricatum.*

Parrotfishes are a large tropical family with representatives in all oceans. They are placed in 2 subfamilies, spp. of one with numerous conical teeth, the other with teeth fused into beak-like plates, similar to parrots, hence their common name. In total 9 genera, about 80 species, the majority of which feed by scraping algae off dead coral rock or rubble and sand. In the process much of the coral surface is crushed and digested. In this way they produce vast quantities of coral sand. Abundant on coral reefs, they are an important component in the reef-building process. Most species have differently coloured juvenile stages and sexes. Males are often colourful, females are rather drab, their identification sometimes only being possible by observing interaction with males.

Scarus strongylocephalus **Sumatra**

Indian parrotfish

Le: To 70 cm. Di: Southwest Indonesia. De: 2-35 m. Ge: Common, widely distributed throughout Indian Ocean. Singly or in pairs in lagoons and on outer reefs, rarely also in schools. Juveniles are usually seen singly. Left: Male. Below: Female.

Dusky parrotfish

Le: To 35 cm. Di: Entire area. De: 3-40 m. Ge: Coastal, often silty habitats, on crests and drop-offs and protected outer reef areas. Right: Male. Below: **Yellow-head parrotfish** *S. spinus*, to 30 cm, trop. W-Pacific, 3-20 m.

Scarus niger **Flores, Indonesia**

Chameleon parrotfish

Le: To 31 cm. Di: W-Pacific. De: 1-30 m. Ge: Female phase drab brown but male has splendid colours, including purple and orange. Both can change coloration. Dark stripe from snout below eye with 2nd parallel stripe distinct. Below: **Orange-blotch parrotfish** *S. bowersi*, to 40 cm, Java to S-Japan, 3-30 m.

Scarus chameleon **Yonaguni I., Japan**

Two-colour parrotfish

Le: To 90 cm. Di: Tropical West Pacific. De: 1-40 m. Ge: Adults (right) are rather shy and look very different from the juvenile (small photo below). Distinct juveniles are often noticed by divers and live singly while adults congregate into groups usually led by a very colourful male.

Cetoscarus bicolor **Hong Kong**

SAND DIVERS — TRICHONOTIDAE

Trichonotus setigerus **Alor I., Indonesia**

Blue-spotted sand diver
Le: To 22 cm. Di: Widespread tropical Indo-Pacific. De: 3-40 m. Ge: On clear coastal sand flats and slopes, in small loose groups, often mixed with next species. Usually perched with ventral fins on sand, waiting for prey. Below: **Long-rayed sand diver** *T. elegans*, to 18 cm, Indonesia to S-Japan, 1-40 m.

SANDPERCHES — PINGUIPEDIDAE

Parapercis schauinslandi **Kerama I., Japan**

Lyre-tail sandperch
Le: To 18 cm. Di: Indonesia to Philippines. De: 10-50+ m. Ge: In small groups on coastal, current-prone slopes on sand, feeds high above substrate on plankton. Below: **Thousand-spot sandperch** *P. millepunctata,* 18 cm, entire area, 2-15 m.

Parapercis clathrata **Bali, Indonesia**

False-eye sandperch
Le: To 24 cm. Di: Eastern Indonesia to Philippines. De: 2-15 m. Ge: On coastal and outer reef crests and rubble slopes. Similar to **Black-banded sandperch** *P. tetracantha* (below), to 23 cm, entire area, 1-25 m, bold black and white patterns and yellow on lower lip distinct.

A very large and complex family of mostly very small species, with more than 50 genera and 300+ species, most of which are distributed in tropical seas. They feature a slimy skin without scales, the teeth consist of a comb-like arrangement in each jaw, the canines are greatly enlarged in certain species. Various subfamilies.

Some blennies mimic other fishes, including other blennies. The best-known is the impersonator of the cleaner wrasse *Labroides dimidiatus*. Primarily benthic, but some specialised species feed on zooplankton, others attack other fishes well above the substrate and feed on their external parts, approaching them with trickery, such as mimicking harmless species. The bottom-dwelling species feed on a mixed diet of algae and invertebrates, those living in intertidal zones primarily feed on algae, those living deeper primarily feed on invertebrates.

Most blennies take refuge in small holes in rocks, discarded worm tubes or shells. Males attract the gravid females by dancing and displaying colours. Eggs are deposited inside the hole, and guarded by the male. After a few weeks the tiny transparent hatchlings (about 2 mm long) swim to the surface and are taken out to sea by the currents. Postlarvae settle at the bottom when about 10 mm long.

Snake blenny
Le: To 40 cm. Di: Widespread, entire area, ranging to subtropical zones. De: 1-50 m. Ge: Coastal sand or mud flats. Lives in vertical holes made in the mud by other animals. Usually only head exposed, occasionally swims to another hole. With the blenny-typical eel-like body, retreats into hole tail-first.

Xiphasia setifer **Lembeh Strait, Indonesia**

Scale-eating fang blenny
Le: To 14 cm. Di: Entire area. De: 1-20 m. Ge: Sneaks up on unsuspecting fish or diver well above the bottom. Apparently finds human skin as palatable as the pieces of skin, fins, scales or mucus of other fishes that it normally eats. Below: **Lance blenny** *Aspidontus dussumieri*, to 12 cm, entire area, 3-20 m.

Plagiotremus tapeinosoma **Bunaken I., Indonesia**

Meiacanthus atrodorsalis **Bali, Indonesia**

Eyelash fang blenny
Le: To 10 cm. Di: Entire area. De: 5-20 m. Ge: On coastal coral-rich shallow reefs. Two long canine teeth (fangs) in the lower jaw with venom glands at bases, used in defence only. Below: **Smith's fang blenny** *M. smithi*, to 7.5 cm, tropical eastern Indian Ocean to Java, Indonesia (photo), 1-20 m.

Meiacanthus ditrema **Irian Jaya**

Schooling fang blenny
Le: To 6.5 cm. Di: Eastern Indonesia to Philippines. De: 3-25 m. Ge: Clear coastal reefs in rich current-prone areas where they form large schools to feed on zooplankton. Below: **White-lined fang blenny** *M. abditus*, to 8 cm, Indonesia, Philippines, 2-15 m.

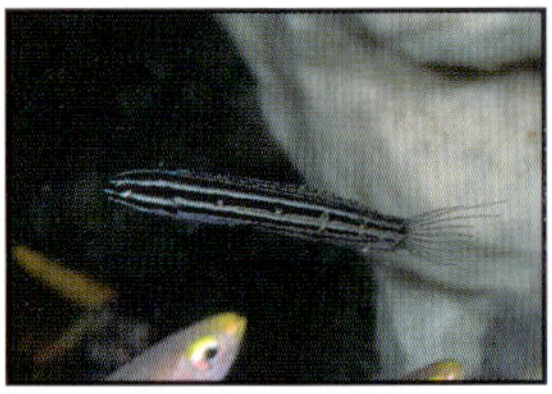

Petroscirtes breviceps **Flores, Indonesia**

Shorthead fang blenny
Le: To 12 cm. Di: Entire area. De: 1-40 m. Ge: On coral reefs, mimics and swims with *Meiacanthus grammistes* (photo: lower) or similar spp. Large canines without venom. Below: **Xestus fang blenny** *P. xestus*, to 6.5 cm, entire area, 1-20 m. Adults use empty mollusc shells for home and nesting.

Sea star guard

Le: To 4 cm. Di: Indonesia to Taiwan. De: 2-25 m. Ge: Likes to perch on sponges, allows close approach. Typical blenny features of short tentacles on forehead, obviously different from gobies.
Below: **White-lined blenny** *E. pictus,* to 4.5 cm, Indonesia to Philippines, 15-40 m.

Ecsenius axelrodi **Batangas, Philippines**

Anemone guard

Le: To 4 cm. Di: Southern Island chain of Indonesia from Bali to Banda Sea. De: 3-30 m. Ge: On clear coastal and protected inner reefs in invertebrate-rich areas, on reef crests and drop-offs to moderate depths. Right: Male. Below: Female.

Ecsenius bathi **Alor I., Indonesia**

Yaeyama blenny

Le: To 6 cm. Di: Our area north to Taiwan and Yaeyama Islands, Japan. De: 0-15 m. Ge: On rocks and living corals in shallow coastal reefs.
Below: **Yellow-eyed blenny** *E. melarchus,* 5.5 cm, Indonesia to Philippines, 3-25 m. In small groups on clear coastal reef crests, in lagoons.

Ecsenius yaeyamaensis **Okinawa, Japan**

Salarias segmentatus **Palau**

Red-spotted blenny
Le: To 11 cm. Di: Indonesia, Flores to Micronesia. De: 4-15 m. Ge: In small loose groups in clear outer reef lagoons, along the bases of rich coral slopes on coral bases with algae on which they graze. Below: **Jewel blenny** *S. fasciatus,* to 14 cm, entire area, 0-6 m. In algae-rich reefs.

Cirripectes stigmaticus **Mindoro, Philippines**

Reticulated blenny
Le: To 10 cm. Di: Tropical West Pacific. De: 0.5-8 m. Ge: Among *Acropora* and *Pocillopora* corals on the surf-swept algal ridge. Feeds on small benthic invertebrates and algae. Below: **Earspot blenny** *C. auritus,* to 9 cm, our area north to Philippines, 1-20 m. On coral heads in shallow outer reefs.

Exallias brevis **Banda Sea, Indonesia**

Leopard blenny
Le: To 14 cm. Di: Widespread in tropical West Pacific. De: 1-15 m. Ge: Clear coastal reefs with mixed coral and algae, in sheltered lagoons on coral and rock. Often common, secretive and usually shy, hiding in corals when approached. Coloration varies from yellow to red. Eggs are laid on dead coral patches.

Mandarinfish

Le: To 6 cm. Di: Indonesia and New Guinea to Philippines and Japan. De: 3-30 m. Ge: Locally common but difficult to find. In dense coral heads. Coastal, often silty habitats, in corals or on open rubble with thin algae.

The dragonet family is large, comprises at least 9 genera and about 125 species, and has numerous small species in the tropics. Fishes with broad spiny heads, instead of scales a tough slimy skin which is of bad taste and with a strong odour, giving some species the common name of stinkfish. Species are benthic, many bury in sand, the reef dwellers hug the substrate closely. During spawning the pair rise slowly from the ground with ventral fins kept together. Eggs float, larvae are planktonic. Adults range from very shallow depths to 400 m, depending on the genus or species.

Centre right: Mating pair. See also *S. moyeri* in THE SENSUAL SEA on p.26-31.

Below: **Picture dragonet** *S. picturatus,* to 7 cm, Philippines and a few Indonesian localities, 1-6 m. Shallow coastal reefs.

Synchiropus splendidus **Gulf of Thailand**

Synchiropus splendidus **Lembeh Strait, Indonesia**

Bartels' dragonet

Le: To 4.5 cm. Di: Eastern Indonesia, Flores, to Philippines. De: 3-35 m. Ge: At Flores fairly common on coastal reef crests, on coral heads, or on rubble patches. With series of white spots along lower sides, adults with tiny blue ocelli on back.

Synchiropus bartelsi **Komodo I., Indonesia**

Synchiropus moyeri **Bali, Indonesia**

Moyer's dragonet

Le: To 7.5 cm. Di: Indonesia and Philippines to Japan. De: 6-50 m. Ge: Very similar to *S. ocellatus* (see centre small photo) but with red bars on anal fin. Rocky coastal reef flats and slopes with short algae and sparse coral growth. Left: Male. Below: Female.

Synchiropus ijimai **Iriomote I., Japan**

Japanese dragonet

Le: To 10 cm. Di: S-Japan. De: 10-20 m. Ge: On rocks with algae. Left: Male. Below: Male **Marbled dragonet** *S. ocellatus,* to 65 mm, Indonesia, Malaysia and Philippines to Japan, Intertidal to 50+ m. 1st dorsal fin of male very tall. Shallow coral reef flats and rocky areas in sheltered coastal bays.

Dactylopus dactylopus **Nabucco I., Indonesia**

Fingered dragonet

Le: To 18 cm. Di: Widespread in tropical West Pacific, entire area. De: 1-50 m. Ge: Sand and mud flats or slopes from very shallow depths to deep mud flats. Sometimes buries during the day and raises fins when disturbed before dashing off. Left: Male. Below: Female.

SPLIT-TONGUED SEA SNAKES

It is common that sea snakes are extremely poisonous but harmless to people. And yet, cases in which they attacked have been reported. The author has frequently met the sometimes scary coral reef dwellers while diving in South East Asia. Therefore he has thoroughly studied these split-tongued underwater reptiles.

HOWARD HALL

Only using diving goggles a Filipino catches a sea snake *(Emydocephalus annulatus)* by hand.

We had been just getting ready to dive when we noticed a wooden outrigger boat of the local natives which anchored not far from our position. On board were two men one of whom was staring intensely at the water's surface. It wasn't very long before he took several deep breaths, glided into the water from the side of the boat and dived down. He surfaced again after a minute. We became curious and dived in the direction of the outrigger boat.

While we swam along the bottom at a depth of 20 m, the Filipino quickly passed us by! He wasn't even wearing fins. His equipment only consisted of wooden goggles which were strapped to his head by a rubber band. With calm, strong strokes of his arms he approached the seafloor where a 1.5-metre-long sea snake was lying. Using his left hand he grabbed the reptile right behind the head. The snake was hardly opposing his grip. He held it between his fourth and fifth fingers while he swam further. The next snake he put between his third and fourth fingers, the third snake between second and third fingers. Finally he surfaced with five snakes in his left hand! We continued to watch and photograph his strenuous work for the rest of the day and were amazed that he was not bitten. Later we heard that he was a master of his trade and was paid a few pennies for each snake he sold.

Sea snakes are hunted for their beautiful skins in the Philippines and for their savoury meat in China. At the beginning of the 1970s it became fashionable to wear shoes, belts and handbags made of the elegant, light-weight but yet tough snake leather. During the following years the Philippines exported large amounts of snake leather products to North America and Europe. The great demand for the precious ware rapidly led to a dramatic decrease of the sea snake populations in those areas where they were preferably collected. The fishermen knew that the animals were present in large numbers during their mating season and took advantage of this fact. In 1974 alone they caught over half a million adult sea snakes around the small Philippine island of Gato! But these times are gone forever. Today sea snakes only rarely stray into this region.

Having witnessed the capturing of sea snakes and having had several harmless underwater encounters with those reptiles, I thought for a long time that they were not aggressive at all. But after an incident at Ashmore Reef in the north-west of Australia I changed my mind. We intended to move a 150-litre-gasoline barrel from shore to a boat anchored 100 m away across the reef flat. We had tied ropes to the floating barrel and then started pulling it stroke after stroke towards the boat. This action obviously raised the interest of a sea snake gliding past us. It changed direction and climbed onto the rolling barrel. We were absolutely amazed as we watched the sea snake permanently trying to bite into the barrel! There were no doubts left about the

The split tongue of the Olive sea snake *(Aipysurus laevis)*.

snake's aggressiveness because even when we hoisted the barrel onto the deck of the boat the snake bit it again and again. So much about the theory of harmless sea snakes!

Admittedly sea snakes are rarely seen by divers or reef waders on most of the tropical shores. One has to distinguish between the amphibious sea kraits of the subfamily Laticaudinae which are very common in our region, and the true sea snakes of the subfamily Hydrophiinae which are truly marine and never leave the (sea) water. All species of the latter subfamily inhabit the tropical Indo-Pacific Ocean from the Arabian Gulf to Malaysia and Indonesia, to the Philippines and the Coral Sea. As an exception the Pelagic sea snake *Pelamis platurus* has an even wider distribution, occurring at the coast of East Africa as well as on the Pacific coast of the Americas. It is even found at the surface of the open ocean, sometimes hundreds of kilometres away from the continent. Often several of these snakes entwine each other to form regular rafts. In the Atlantic Ocean - and hence also in the Caribbean -, sea snakes have not been found yet. In the Indonesian Banda Sea the local fishermen refer to the island of Manuk as the capital of sea snakes. I myself, however, have encountered only 4 species underwater at this island, including the rare sea krait *Laticauda semifasciata*. Excellently camouflaged and leading a secretive life these snaking sea creatures can be found everywhere in South East Asia. Most sea snakes are helpless on dry land and unable to bite. Despite this fact they shouldn't be touched because their venom is very potent. The most poisonous of all species is *Enhydrina schistosa* which is distributed from the Seychelles to the Coral Sea. It can produce 10-15 milligrams of a toxin of which a mere 1.5 milligrams can kill a human adult. On the other hand, there are no confirmed reports of unprovoked fatal attacks on divers, snorkellers or reef waders. Only from Southeast Asia there are reports about fatal accidents with sea snake catchers, who were bitten while removing the snakes from their nets.

PHOTOS: HELMUT DEBELIUS

A dive master demonstrates how dangerous sea kraits really are. Manuk Island, Banda Sea, Indonesia.

Above and right: Sea krait (*Laticauda semifasciata*) resting in the reef and swimming freely about at Manuk Island.

Sea snakes mainly feed on benthic fishes like gobies and garden eels. Sea kraits are diurnal hunters and with some luck you may see them in caves or cracks of the reef. When my son was 6 years old, we visited Bali and were exceptionally fortunate. Before entering tropical waters for the

Sea kraits *(Laticauda colubrina)* are not strictly marine. Here several of them are resting among the roots of mangroves. Malaysia.

first time to snorkel there he had already practised in the deep end of the public swimming pool back home. On the morning after our arrival on Bali we intended to do the real thing and donned face mask and snorkel. Directly in front of the hotel there was a large breakwater and the surface of the water behind it was absolutely calm. Hoping to find some coral fish in the nooks and crannies of the breakwater we snorkelled right away. My son started to jubilate when he saw some turquoise blue damselfishes a few metres in front of the breakwater.

Always concerned with his safety, I was watching each and every one of his movements when he suddenly pointed down. At a depth of only 2 metres a sea krait *Laticauda colubrina* about 70 cm in length made its way among the coral rubble. I was immediately able to identify the species by its characteristic head pattern. At the same time I noticed an even smaller free swimming moray. Funny enough, because normally only the heads of these fishes are visible, peering from the cracks they hide in. Additionally, this juvenile moray moved in a peculiar and totally untypical way. I swam down to pick it up and realised that this moray was paralysed. Now I understood: the sea krait must have bitten the fish to devour the stunned victim later on. This is exactly how most sea snakes and sea kraits hunt. Slowly 'our' sea krait circled the moray while still peeking into each hole in the ground without any sign of excitement. Again I went down to pick up the moray only to drop it onto the sea krait. Nothing happened. While the moray was hardly moving, I tried to explain to my son what was going on there. On the one hand I desperately wanted to fetch my underwater camera, still stored in the suitcase, from the hotel room; on the other I couldn't leave my son alone during his first snorkelling venture! After another 2 minutes the sea krait suddenly started to pace up: Its tail wiggled violently about as soon as the snake touched the motionless body of the moray with its head. Fascinated we watched from

The main occupation of Sea kraits *(L. colubrina)* under water is hunting. Malaysia.

above how the reptile bit into the moray's dorsal fin only to release it and move its head up and down along the moray's long body like playing a mouth organ. Obviously the snake was searching for the victim's head: having reached the end of the tail it immediately returned along the body. Finally it found the head, bit into it and started to devour the moray as it twitched convulsively. We clearly saw the fish being moved inside the sea krait towards its posterior end. I was silently cursing along because such a scene couldn't have been easier to photograph. But then my son had been witness to something many experienced divers dream of for a lifetime!

MARK PREIN

The rare Sea krait *Laticauda laticauda* crossing a coral reef in the Philippines in search of prey.

Reptiles reproduce in two ways: Some like the sea kraits lay eggs and are hence called oviparous. Others like the sea snakes give birth to live young and are hence called viviparous. The latter give birth to already fully developed young underwater. Immediately after being born the young swim directly to the surface and draw their first breath. At the same time this is their first step towards independence from their mother, who like all sea snakes does not take care of her offspring. Compared to viviparous terrestrial snakes, sea snakes give birth to less young: on average only about a dozen while the litter size of their terrestrial relatives is often more than 40. Also interesting is the arrangement of young inside the mother's body, which are situated much further towards its front when compared to terrestrial snakes. Most probably the reason for this is keeping the unborn away from the mother's paddle-shaped tail which otherwise would be hindered in its movements necessary for swimming. In most species the gestation period of sea snakes is about double that of life-bearing terrestrial snakes, namely 8-9 months.

Sea kraits deposit litters of various sizes on dry land: *Laticauda colubrina* on Palau had 4-10 eggs, while the same species had 12-14 eggs in New Caledonia. Oviparous terrestrial snakes deposit about 45 eggs. The reproduction phases of marine snakes are totally different from each other. Some species pair throughout the year, others only during a certain period. The Olive sea snake *Aipysurus laevis* is even known not to reproduce at all in some years.

The size of adult specimens from both families of marine snakes varies considerably. Some grow to only half a metre, others to almost two metres in length. Sexes can be separated only by size, the females being bigger. The coloration is the same in both sexes. The growth of *Enhydrina schistosa* has been investigated: females become sexually mature at the age of 18 months, however, only 6% of all females survive to this age and only very few specimens reach an age of 4 years. *Aipysurus laevis* is known to reach an age of 10 years.

Like their terrestrial counterparts sea snakes breathe air and therefore have to return to the surface at regular intervals. This is probably their only handicap in the ecological niche, the coral reef, which they inhabit. Hardly any information is available on the duration of their dives. It is the opinion of some experts that they are able to stay underwater for hours without breathing. I have observed

While the sea snakes' hide is processed into fine shoes in the Philippines, their meat is being eaten in China.

Close-up of a resting sea snake *(Astrotia stokesi)* from Ambon, Indonesia.

individual sea kraits at Manuk in the Banda Sea which swam to the surface once or twice during my 50-minute-dive. Without effort and almost leisurely they wiggled their paddle-like tail and - without caring about the photographing diver - made their way up towards the sun which was high above the horizon. At the surface they drew a short breath and went down again into the reef. I didn't observe any of the timid movements usually seen in fishes when divers approach them too closely. Also, before and after breathing at the surface, the snakes swam along the reef without seeking any shelter as if there were no hungry predators to worry about. We saw sea kraits at a depth of 40 m, but most marine snakes live or at least hunt between 5 and 10 m as can be deducted from the fishes found in their stomachs, which are from the shallow reef areas.

Venomous animals are assumed to have only few enemies. This is also true for many sea snakes which are avoided by potential predators in the reef. But sea eagles and tiger sharks - their greatest enemies - are not to be scared away. The Pelagic sea snake *Pelamis platurus* clearly signals its ability to defend itself. It is contrastingly coloured and the yellow colour of its belly certainly advertises its potential toxicity. Experiments in large aquaria have shown that large groupers as well as hungry White-tip reef sharks keep at a safe distance from this species. Even the cut-up pieces of this sea snake that were presented to these predators were avoided. The fishes clearly responded to more than only visual experience. These predatory fishes all came from the Pacific coast of Panama. After the first experiments the predators were subsequently caught on the Caribbean side of Panama where there are no sea snakes. One of the Pelagic sea snakes was immediately attacked and eaten by a large snapper. But the snapper soon spat out the snake and died shortly thereafter. *Pelamis* defended itself against other attackers and killed two more Caribbean predatory fishes by biting them. Until then these fishes had only been familiar with similarly looking Caribbean snake eels.

Eels, especially those of the family Ophichthidae (snake eels), are ever and again taken for sea snakes. The colour pattern of *Myrichthys colubrinus* is almost identical to that of most sea kraits. However, when taking a closer look one can easily notice the differences: sea snakes and sea kraits have distinctly large scales and a paddle-shaped tail. Eels have no scales, a long dorsal fin and of course - being fishes after all - gill openings behind their head. The largest sea snakes reach a maximum length of 2 m while snake eels grow to only half that length.

There is no need to panic for a diver who suddenly finds himself face to face with a snake-like creature underwater. Snake eels certainly flee at once and sea snakes are best left alone. It is out of the question that imitating the Philippine snake catchers and trying to touch them with an unprotected hand is not a good idea. Often one finds them at rest and whenever they are swimming around a diver should simply change his direction.

The distinctly patterned Pelagic sea snake *(Pelamis platurus)* has the widest range of distribution of all sea snake species.

The largest family of marine fishes, comprising over 200 genera, and an estimated 1,500 species worldwide. The majority of species are tropical, and live on reefs or nearby on sand. Gobies are primarily benthic marine fishes, usually keeping in touch with the substrate, a few readily hover above the ground. Many of the sand gobies completely bury in the sand at night, others use burrows as shelters. Some live in brackish water and enter the lower reaches of streams, a few are totally adapted to freshwater. Because of their small size and secretive habits, many new species are being discovered on a regular basis, especially in the Asian Pacific region, and the total number of species is even difficult to guess at present. Many species are highly specialised and associate with various invertebrates, ranging from certain coral species and sponges to crustaceans, especially members of the family Alpheidae, the distinctly patterned and often colourful partner shrimps.

Most gobies pair to breed and use burrows, narrow ledges, empty shells or any small enclosure for a nesting site. They produce several hundreds of sticky eggs which are laid in a neat patch. The male or both sexes guard the eggs and the nesting site. Hatching larvae are attracted by light and swim quickly to the surface.

Further gobies can be seen in the Stories TINY FISHES - A PSYCHIATRIST'S VIEW on p.102-105 and TWO SHARP EYES WILL EARN YOU A COZY HOME on p.244-247.

Luposicya lupus **Sipadan I., Malaysia**

Wolf goby

Length: To 4 m.
Distribution: Malaysia, Indonesia, Philippines.
Depth: 1-15 m.
General: Usually found sitting on the mantle of giant clams of the genus *Tridacna,* e.g. *T. gigas.*

Below: **Soft-coral goby** *Pleurosicya boldinghi*, to 4.5 cm, entire area north to Japan, 20-70+ m. A stocky, white-looking goby found on soft corals, especially large *Dendronephthya* spp. and usually in small groups. On deep coastal slopes with isolated soft coral colonies and usually in depths over 25 m, one specimen known from a 130-m-trawl.

Many-host goby
Le: To 3 cm. Di: Widespread tropical Indo-Pacific. De: 2-20 m. Ge: On various invertebrate hosts such as sponges, soft or hard corals, plants, and ascidians (photo right). Below: **Slender sponge goby** *P. elongata,* to 4 cm, Indonesia, New Guinea, probably widespread, 3-50 m. Underside of fan sponges.

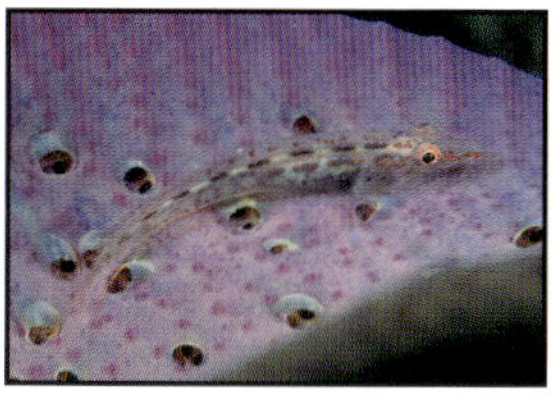

Pleurosicya mossambica **Bali, Indonesia**

Loki whip goby
Le: To 3 cm. Di: Widespread, common in Indonesia and Philippines. De: 3-30 m. Ge: In pairs or small groups on smooth whip corals and gorgonians. Below: **Hovering goby** *B. natans,* to 2.5 cm, entire area, 7-27 m. In groups hovering just above the branches of *Acropora* corals.

Bryaninops loki **Lembeh Strait, Indonesia**

Graceful partner goby
Le: To 5 cm. Di: Area north to Ryukyus. De: 2-20 m. Ge: Pectoral fins large, fan-like, constantly waved while hovering above burrow in sand on reefs. Very shy, associated with burrowing shrimp *Alpheus rubromaculatus*. Below: **Crab-eyed goby** *Signigobius biocellatus,* 10 cm, north to Philippines, 6-30 m.

Lotilia graciliosa **Derawan I., Indonesia**

Valenciennea puellaris **Bali, Indonesia**

Orange-spotted sleeper goby
Le: To 14 cm. Di: Entire area. De: 3-25 m. Ge: Coastal reef slopes and inner reefs, usually in rubble zones, juveniles in small aggregations, adults in pairs. Below: **Broad-barred sleeper goby** *V. wardi,* to 12 cm, entire area, 10-30 m. Very rare, on coastal slopes, on sand or silt.

Valenciennea strigata **Sangihe I., Indonesia**

Blue-streak sleeper goby
Le: To 18 cm. Di: Our area north to Ryukyus. De: 1-20 m. Ge: Common. In clear lagoons and on outer reefs. On hard substrates as well as on sand or rubble. Adults usually in pairs, juveniles in small groups. Below: **Green-band sleeper goby** *V. randalli,* to 16 cm, all of area, 10-25 m. Mud or sand.

Amblygobius phalaena **Mindoro, Philippines**

Banded goby
Le: To 15 cm. Di: Widespread in tropical West Pacific, Indonesia to Philippines. De: 1-20 m. Ge: On sand, rubble, leaf debris. In pairs, burrow often under a rock. Filters small invertebrates and algae from sand. Below: **False sleeper goby** *A. nocturnus,* 7-10 cm, entire area, 1-30 m.

Pink-lined goby
Le: To 10 cm. Di: Entire area. De: 3-30 m. Ge: Coastal and inner reefs, lagoons and slopes. On coarse rubble and low reef in sandy areas. Juveniles singly, adults in pairs, in depths of about 10 m. Below: **Byno goby** *A. bynoensis*, to 10 cm, Indonesia, Java, 1-10 m. Clear inshore lagoons and protected bays.

Amblygobius decussatus **Kerama I., Japan**

Rainford's goby
Le: To 8.5 cm. Di: W-Pacific. De: 3-30 m. Ge: Clear coastal reefs and protected bays with rich coral and algae growth, slopes and shallow drop-offs, in areas subject to moderate currents. Below: Similar **Hector's goby** *A. hectori,* to 8 cm, our area along Asian mainland to southern Japan, 5-20 m.

Amblygobius rainfordi **Batangas, Philippines**

Black-ray partner goby
Le: To 6 cm. Di: Indonesia, Malaysia, Philippines. De: 8-40 m. Ge: Coastal sand flats, slopes with sparse rubble, in current-prone areas. Adults in pairs, hovering above burrow, waiting for zooplankton. With *Alpheus randalli.* Below: **Yellownose partner goby** *S. xanthorhinica,* 6 cm, area to S-Japan, 3-45 m.

Stonogobiops nematodes **Mabul I., Malaysia**

Mahidolia mystacina **Flores, Indonesia**

Smiling partner goby
Le: To 8 cm. Di: Entire area. De: 6-30 m. Ge: Common, variable from bright yellow to brown or black. First dorsal elongated in males. Singly or in pairs in protected bays on mixed sand and mud substrates with a grey-brown alpheid. Below: **Blue partner goby** *Myersina* sp., to 6.5 cm, Indonesia, 10-30 m.

Cryptocentrus cyanotaenia **Mabul I., Malaysia**

Lagoon partner goby
Le: To 12 cm. Di: Indonesia. De: 1-6 m. Ge: Well protected lagoons with intertidal reef flats, from silty coastal to clear outer reefs, very shallow. In pairs, with black alpheid, often many pairs present in small areas. Below: **Spot-face partner goby** *C. polyophthalmus,* to 15 cm, Indonesia, 10-40 m.

Amblyeleotris wheeleri **Palawan, Philippines**

Wheeler's partner goby
Le: To 10 cm. Di: Entire area. De: 5-15 m. Ge: Widespread also in the Indian Ocean. Prefers clear lagoons and seaward reefs.
Below: **Singapore partner goby** *Cryptocentrus singapurensis,* to 10 cm, Indonesia, Singapore, Malaysia to southern Japan, 1-10 m.

Magnificent partner goby
Le: To 12 cm. Di: Malaysia to S-Japan. De: 15-40 m. Ge: Known for many years this most beautiful partner goby with its massive 1st dorsal fin is still undescribed. Shown here with *Alpheus randalli.* Below: Another member of this group, the **Fan partner goby** *F. latruncularius.*

***Flabelligobius* sp.** **Derawan I., Indonesia**

Sailfin partner goby
Le: To 12 cm. Di: Indonesia to S-Japan. De: 10-50 m. Ge: On sand in caves or below large overhangs along drop-offs on reefs. Singly or in pairs. Lives together with different snapping shrimps. Sail-like 1st dorsal and coloration distinct. Below: **Metallic partner goby** *A. latifasciata,* 13 cm, Indonesia, 10-40 m.

Amblyeleotris randalli **Ambon, Indonesia**

Spotted partner goby
Length: To 8 cm. Distribution: Entire area. Depth: 8-30 m. General: A common but very attractive partner goby easily recognised by its orange spots and dark patch on belly. As shown here the blind snapping shrimp *Alpheus ochrostriatus* maintains contact with the goby with one of its antennae. If danger approaches, the goby signals the shrimp with a flick of its tail and the shrimp immediately retires to safety down into the burrow. More shrimp-goby partnerships are shown in TWO SHARP EYES WILL EARN YOU A COZY HOME on p.244-247.

Amblyeleotris guttata **Cebu, Philippines**

Amblyeleotris yanoi **Bali, Indonesia**

Yano's partner goby

Le: To 11 cm. Di: Entire area. De: 5-20 m. Ge: Usually singly with 1 or 2 brown and white snapping shrimps. Below: **Red-margin partner goby** *A. gymnocephala,* to 12 cm, Indonesia, 6-40 m. Thin red outer margin of whitish median fins distinct. Coastal white to dark sand and mud slopes, singly or in pairs.

Asterropteryx ensifera **Mabul I., Malaysia**

Blue-speckled goby

Le: To 3.5 cm. Di: Indonesia to S-Japan. De: 6-40 m. Ge: Common, on bare rubble slopes, usually in large numbers, hovering just above the substrate to feed on zooplankton. Coastal to inner reefs in current-prone areas. Below: **Poisonous goby** *Yongeichthys nebulosus,* to 15 cm, entire area, 6-30 m. Skin toxic.

Exyrias puntang **Bunaken I., Indonesia**

Puntang goby

Le: To 14 cm. Di: West Pacific. De: 1-10 m. Ge: Shallow coastal lagoons and rock pools, often in freshwater run-offs. On rocks or coral rubble, usually seen singly. Best recognised by the spotted pattern in the first dorsal fin. Below: **Mud-reef goby** *E. belissimus,* to 15 cm, entire area, 3-20 m.

Lemon coral goby

Length: To 6.5 cm.
Distribution: Entire area.
Depth: 5-25 m.
General: Yellow with thin blue lines. Body stout when compared to other species, unmistakeable by coloration and habitat: Found exclusively in heads of branching *Acropora* corals. Common in lagoons, sometimes forming colonies. Like its congeners it produces a large amount of thick, sticky mucus that has a strongly bitter taste and probably is used to deter predators.

Gobiodon citrinus **Singapore**

Barred coral goby

Le: To 3.5 cm. Di: Entire area. De: 3-25 m. Ge: *Gobiodon* spp. live on branches of living coral, have no scales and are covered with a poisonous mucus. They are able to resist capture by wedging themselves tightly in the coral branches. Below: *Gobiodon* sp., known for years from our area but still undescribed.

Gobiodon histrio **Halmahera, Indonesia**

Okinawa coral goby

Le: To 3.5 cm. Di: Our area north to Taiwan and Okinawa Islands, Japan. De: 3-20 m. Ge: On *Acropora* corals, usually in groups of up to 15 individuals. Bright yellow but little noticed. Below: **Dinah's goby** *Lubricogobius* sp., 3 cm, PNG, rest uncertain, 5-35 m. On sponges, enters openings if threatened.

Gobiodon okinawae **Kerama I., Japan**

Nematelotris helfrichi **Okinawa, Japan**

Nemateleotris helfrichi **Palau**

Helfrich's fire goby

Length: To 6 cm.
Distribution: Ryukyu Islands to Palau and eastward. Depth: 25-70 m but rare above 40 m.
General: Hovers near its hole in the ground, seen over patches of sand, rubble or hard substrates near steep outer reefs. Usually living in pairs but these are difficult to photograph together because they are very shy.

Dart gobies are a moderately large family comprising 12 genera and about 45 species placed in two subfamilies: Microdesmidae (worm gobies) and Ptereleotrinae (dart gobies). All have large eyes and prominent dorsal and anal fins. Often in burrows made by other fishes.

Below: **Red fire goby** *N. magnifica,* to 8 cm, entire area, 6-61 m. Found over patches of sand, rubble, or hard open bottom at the bases of reefs. It hovers above the bottom to feed on zooplankton and darts into its hole when alarmed. Adults inhabit the hole in pairs, juveniles are seen in groups in caves and crevices.

Nemateleotris decora **Batangas, Philippines**

Purple fire goby

Le: To 9 cm. Di: Entire area. De: 27-70 m. Ge: Deep, over rubble and coral blocks as well as sand or fine rubble. Hovers within a half meter of the bottom, facing into current to feed on plankton. Very shy. This goby has been observed in the same cave as the *Pseudochromis porphyreus* in the background of the photo.

Blue-barred ribbon goby

Length: To 20 cm.
Distribution: Indonesia, Philippines.
Depth: 20-40 m.
General: Lives in burrows on mud or fine silt bottom. Adults in pairs seen hovering tail-down above burrow. Juveniles in small groups, always close to burrow. Feed on passing zooplankton. Burrows are slot-shaped which matches the flattened, ribbon-like bodies of these gobies. Little-known, probably because of for divers unattractive habitat but beautiful and interesting fish.

Oxymetopon cyanoctenosum **Flores, Indonesia**

Lined dart goby

Length: To 10 cm.
Distribution: Entire area.
Depth: 35-50+ m.
General: Rarely seen shallower than 45 m, this dart goby is usually found on deep outer reef slopes where the slope flattens to a sand ledge. There are typically currents over the ledge. Probably widespread but not reported because of the depth. Feeds on zooplankton and swims above the bottom but will retreat into burrow if threatened.

Ptereleotris grammica **Yonaguni I., Japan**

Zebra dart goby

Le: To 11 cm. Di: Entire area. De: 2-25 m. Ge: It typically forms large aggregations over hard, gently sloping bottoms, and is probably the wariest species in the genus. Usually small groups occupy the same burrow. Below: **Blackfin dart goby** *P. evides*, 14 cm, entire area, 2-15 m.

Ptereleotris zebra **Banda Sea, Indonesia**

Ptereleotris monoptera **Sulawesi**

Lyre-tail dart goby

Length: To 12 cm.
Distribution: West Pacific, Indonesia, Philippines to New Guinea and warm-temperate Japan.
Depth: 3-50 m.
General: Coloration variable, greenish blue to bright blue, with pinkish anal fin. Dorsal fin continuous, caudal fin lunate, both are diagnostic features. Coastal to clear outer drop-offs, in areas prone to tidal currents. Usually schooling in small to large numbers, sometimes found mixed with other similar species.

Ptereleotris heteroptera **Mabul I., Malaysia**

Spottail dart goby

Le: To 12 cm. Di: Entire area. De: 7-46 m. Ge: Common, on open sand and coral rubble near reefs in 20+ m. Black spot on tail distinct but tail yellow or blue. Often in pairs, retreat into burrow when approached. Below: **Yellow-stripe worm goby** *Gunnelichthys viridescens,* 7 cm, north to Palau, 5-20 m.

Gunnelichthys monostigma **Cebu, Philippines**

One-spot worm goby

Le: To 11 cm. Di: Entire area. De: 1-20 m. Ge: On open sandy substrates in current-rich areas.
Below: **Brown-stripe worm goby** *G. pleurotaenia,* to 9 cm, entire area, 1-8 m. Found in the same habitats as the previous species.

HERE THERE BE DRAGONS

When an interesting encounter with one of nature's rarer creatures turned into something of an aquatic farce, Douglas David Seifert grasped the photographic opportunity, but kept in mind that when it comes to dragon confrontations, 'it's every man for himself'!

SHIMLOCK/JONES

A Komodo dragon resting on a cliff that overlooks the sea near Rinca Island in the southern Komodo National Park.

A large part of the enjoyment of travelling to far-flung destinations is, as the Australians so brilliantly phrase it, 'having a wander'. There is nothing so tranquil as a remote beach, all to oneself or shared with a few friends. The sensation of walking on sand that bears no other footprints is one of the last solitary experiences remaining in travel on this globe of seven billion (and counting) inhabitants. Of course, it is a fiction, really, all in the mind.

Be that as it may, it was still good to get off the boat after a week of diving at Komodo Island, Indonesia. We were walking on Komodo's smaller, neighbouring island, Rinca, when footprints that were distinctly non-human were identified. The shape of the prints bore jagged fore edges, which indicated claws, and a curved furrow sketched itself in the sand sporadically between the parallel footprints, indicating a tail. It could only be one thing, on a remote island in the Indonesian archipelago: a Komodo dragon or dragons, because once the first set of tracks were noticed, and suspicions aroused, others called themselves to our attention.

The Komodo dragon *(Varanus komodensis)* is a species of monitor lizard and the world's largest living lizard; the single largest recorded specimen of which weighed in at a whopping 165.9 kg. The average size is somewhat smaller, a mere 90 kg, and as far as length is concerned, the males are larger than the females by a third (or 3.10 m versus 2.50 m). Tourists can experience Komodo dragons at various zoos around the world or at Komodo Island itself. Indeed, they are the star attraction for enticing visitors to Indonesia's Komodo Island.

As with so much of the natural world, seeing a Komodo dragon is an experience now common to tens of thousands of day-trippers, but only a few centuries ago Komodo dragons were the reason why regions shown as unexplored on the maps used by ancient mariners bore the caveat: 'Here there be dragons.' Well, here indeed, but where?

The tracks meandered up the beach, parallel to the shore, with a definite drift upwards toward the forest edge ahead, for quite a distance. Pausing to reflect that Komodo dragons are carnivores that have been indicted in the killing and devouring of several humans over the years; that their jaws are filled with jagged, conical teeth, oozing with the slime of at least 60 differing bacterial strains (each

PHOTOS: DOUGLAS D. SEIFERT

Like all scaly reptiles (lizards, snakes) the Komodo dragon has a split tongue which is used to sample the air for scent molecules.

of which can cause a fatal infection of blood poisoning and - in this day and age - gangrene) and knowing that they can run, over short distances, at a speed of 18 km per hour, upon seeing these tracks, there was only one sensible thing to do. We followed them.

The beach was wide and the sea lapped gently at shore. Here and there, a nautilus shell lay bleaching in the bright sunshine. A tangle of mangrove trees had colonised a patch of shoreline ahead; amid rocky boulders, their roots grew skyward from a muddy plain. Small fish splashed in the shallows beneath their network of roots and branches. Languid flies buzzed to land and flit upon rotting flotsam and on to the shore party. They weren't a biting kind of fly and they moved so slowly, that they

In earlier times Komodo dragons had to swim a lot when crossing over from one island to the other. Today, they still can swim!

SHIMLOCK/JONES

From a fish eye's view: a Komodo dragon tucks its legs close to its body and wriggles through the water like a snake.

were not swatted into fly oblivion: a gentle wave nearby sent them back aloft on to their lazy flight.

We walked on, with only the scrunch and crunch of sand and shell to keep cadence. I wasn't sure how well dragons could hear, although I'd read that they have keen hearing, much better than humans. Certainly the dragon, or dragons, would know we were coming. The wind was coming from behind us, carrying our scent ahead. I knew their sense of smell was acute.

Komodo dragons have to have heightened senses. As carnivores preying on wild deer, pigs and smaller Komodo dragons, they are ambush predators that must lie in wait and rely upon startling their prey to capture it. Their mouth slams shut upon the victim like a bear trap and a life-threatening wound is inflicted. If the animal does manage to escape, the bacteria in the dragon's saliva goes to work, infecting the victim and killing it slowly by a process of septic poisoning. Then the dragon merely has to find the carcass and feed upon the carrion. It is believed the dragons evolved to their current size as a function of preying upon two species of dwarf elephants that lived in the region in the distant past. The elephants are gone, the dragon remains, a carnivore - always ready to feed.

Based on previous experience, I could just imagine the dragon ahead, lying low beneath the crook of a fallen tree, blending in perfectly with the muted browns and greys of the forest edge. Its head would be raised, with cold, gleaming, black eyes staring forward, and its long pink, forked tongue flicking forward into the air, literally tasting our scent as molecules borne upon the breeze. It would detect five humans, two of which were female, and three males. All reeking of photographic chemistry or electronic circuitry. I was giddy at the thought of photographic opportunities. I was interrupted from my fantasies by the reality before us, a 3 m Komodo dragon, with a horrific trail of drool hanging from its mouth.

For a vicious, dangerous predator, the Komodo dragon really has the air of the comical about itself. It moves its head in a slightly jerky motion consistent with that of birds. Its black eyes are beady, but then so too are many parakeets and budgies. But what lessens the illusion of utter fearsomeness is the flicking of the tongue. It shoots out like one of those noisy party horns. The tongue unfurls a forked tip, the points of which travel outward to the extent of their length and then the entire tongue retracts with a lightning-quick return. The tips of the tongue seem to be scooping a taste of the air, which is, in fact, what they do. The tongue also functions to somehow measure the distance between itself and what it is interested in.

The Komodo dragon flicked its tongue a few more times, cocked its head once, and seemed to make eye contact. It moved out from under cover of the fallen log and began to claw its way down the beach. The dragon gambolled forward, its body writhing with great exaggeration as it progressed and the flat-sided tail swung to and fro like a supple, but deadly, club. To say the dead do not return to life is not to experience a group of photographers pausing from composing pictures though a viewfinder to their realising that a large predator is advancing upon them with gastronomic intent. In one second, from stationary to rapid flight, the shore party moved in a variety of directions, all the while continuing to make pictures, in the process also kicking up a flurry of sand as they positioned themselves out of the dragon's immediate path.

The sudden activity seemed to confuse the dragon. Much in the way that a polarised school of fish use simultaneous activity to keep a predator from focusing upon an individual, the chaotic moves made by the shore party distracted the dragon from 'investigating' a particular individual. I reloaded my camera. This was to be a normal nature shoot after all, not a snuff film.

The dragon wandered past us and walked down to the water's edge. It paused, and the water washed gently over it its scaly feet. Its gaze seemed to survey the entirety of the scene around it: the

SHIMLOCK/JONES

With its head held high above the water the largest of all lizards appears like a swimming dog when viewed from a boat.

islands on the horizon, the sea itself, the Evening Star at anchor in the bay, the shore party, the long white beach with forest behind, and finally, it seemed to settle upon our aluminium tender a few hundred metres down the beach. The dragon flicked its tongue towards our tender and began to trot through the waves towards our boat.

We stopped composing photographs as we realised it was departing. We followed the dragon, flanking it and running ahead for more photos, leapfrogging each other, trying to capture this remarkable lizard on the beach. Its stride was continuous and fast for an animal so stout and low to the ground. We were sprinting to keep up with it and I was panting and wheezing, as usual. We all stopped at the same moment, incredulous with a new realisation, as we comprehended the dragon's definite agenda. Without so much as a pause, the dragon closed the last few metres between it and the boat and bounded on board. The movements recorded themselves in my brain in that queer, slow-motion detail that always accompanies the extraordinary. The dragon seemed to pick up speed at the last moment and bounded upwards, with a leap of its forelegs, reminiscent of a cheetah. Its tongue thrust out flicking into the air and its black eye blinked, as if winking to us. It claws came to rest on the aluminium gunwales and I could swear I heard the screech of claw scratching upon metal. The dragon paused for a beat and struggled with its own bulk as it pulled itself over the pontoon. The beast upended into the tender with a high flick of its tail, which seemed to hang in the air for an eternity before it cantilevered over and disappeared from view as its body crashed to the boat's deck, obscured from view by the silvery pontoon.

"Unreal." "Outrageous." "Anybody get that on film?" "Yeah, think so." Finally, we slowly realised what had happened. Our skipper had been out fishing in the morning but I had assumed he'd been unsuccessful. Apparently not. He had left his catch in our boat! It was then I noticed that the momentum from the dragon's impact with the boat - and subsequent vaulting into it - had had the effect of loosening the anchor's tenuous grasp on the sandy beach. The tender floated out, bow seaward, with the lapping of the waves. The anchor line dragged behind from the stern, the anchor's tines unable to find purchase on the sandy bottom.

The dragon that inspired a party of nature photographers so much. It is well able to kill and eat mammals as large as a deer.

The dragon came into view again. It raised itself on to a bench and stood with its head held high. A small snapper lay across its jaws. The dragon jerked its head back and flipped the snapper sideways into the back of its mouth. It gnashed a few times as the small fish settled into the back of its mouth and disappeared with a noticeable gulp. You could follow the snapper's progress down the dragon's gullet by watching the lump in its throat undulate towards its torso.

The dragon's black eyes regarded us with curiosity, but it also had the distinct appearance of being confused by the rocking motion which was now occurring beneath its stiff body. The dragon looked around and flicked its tongue. It spent much time looking at the water all around the boat upon which it was drifting away from shore. It alternated between looking at the receding shore and the waters around the boat. No one asked the dragon what it was thinking, though it was clear it was thinking quite a bit about its situation and the water all around it. But the dragon must have gone through a dragon's cognitive checklist and come up with the only solution available to it. It looked around, flicked its tongue towards the beach and vaulted head-first over the side of the tender. The dive lacked any sort of aesthetic form, and defined the term belly-flop. The splash was the kind made by deadweight hitting the water flat.

Immediately, its head thrust out of the water, high, like the way a dog swims. It tucked its legs tucked close to its sides and furiously wriggled through the water, undulating like a snake as it moved, very rapidly, for shore. We had heard that Komodo dragons could swim, at least for short distances. But actually seeing one swim, was something unexpected. It wriggled, wriggled, wriggled, all the while keeping its head out of the water and exerting much energy to get out of the water as fast as possible, like a house cat that finds itself in a swimming pool.

In the long-distant past, when the sea levels were lower and there were not the fierce ocean currents that now rage through the Indonesian Lesser Sunda Islands, it is believed that the ancestors of Komodo dragons swam from island to island as a natural course of their life. In the present day, a swimming Komodo dragon is a rarity. Unlike their reptile cousins, the marine iguana and the saltwater crocodile, they lack the physiology to open their mouths underwater without drowning. It is unknown if they have any significant breath-holding ability. So, although a Komodo dragon may, in time of need, be a reluctant aquatic animal, it really is a case of "Sink or swim!"

Fortunately, the boat had not drifted too far from shore and the Komodo dragon easily reached water shallow enough to stand in before it drowned. The dragon crawled from the water and gambolled up the beach. It paused briefly to regard the shore party and its black eye blinked (a wink?), and it flicked its tongue toward them (in contempt?). It resumed its course, continued up the beach, and disappeared into the underbrush, leaving some fortunate photographers in its wake to wonder at the nature of a Komodo dragon, and to ponder how to get off the dragon's island.

Surgeonfishes are a large circumtropical family, comprising 3 subfamilies. Acanthurinae are the largest subfamily, comprising 4 genera and about 50 species, featuring a single fixed spine on each side of the caudal peduncle, some of which are venomous. The spine is not hinged, as suggested in some of the specific literature, but when used in defense or when fighting, the tail is bent which makes the spine stick out. The subfamily Nasinae comprises about 15 species which are usually placed in a single genus. The species feature 1 or 2 bony plates with spines on each side of the caudal peduncle. The subfamily Prionurinae, the species of which are mostly distributed in temperate waters, comprises only a few species which feature a series of bony peduncular plates.

The diet of surgeonfishes varies between species, some graze algae, others feed on zooplankton, some do both, others filter food from the substrate by sifting sand. The planktivorous species often school in great numbers, whilst benthic feeders either school or pair. The reef-dwelling species can be very colourful, juveniles may differ greatly from the adult in colour as well as shape. The peduncular spines are often surrounded by bright colours to serve as a warning, and are venomous in at least one of the species.

Acanthurus pyroferus **Cebu, Philippines**

Pacific mimic surgeonfish

Length: To 20 cm.
Distribution: West Pacific, entire area.
Depth: 2-40 m.
General: Juveniles mimic various *Centropyge* spp. angelfishes, the model depending on the area. Singly in shallow coastal waters, grazing algae on rocks or sand. The mimic gains some freedom out in the open from predators, which have learned not to hunt the angelfishes because they quickly move about, giving predators little chance to strike and make them look for easier prey. Bottom large photo and two small photos below: Juveniles mimicking *Centropyge flavissimus, C. heraldi,* and *C. vrolikii* respectively.

Acanthurus pyroferus **Bali, Indonesia**

Lined surgeonfish

Length: To 35 cm.
Distribution: Widespread tropical Indo-Pacific.
Depth: 0.5-6 m.
General: Common inhabitant of the surge zone. It is territorial and aggressive with large males controlling well-defined feeding territories and harems of females. Caudal fin lunate, corners greatly extended in adults. Easily identified by its colour pattern, even as small juvenile. Usually in very shallow depths, schooling over reef crests with gutters, rarely deeper than 6 m. Juveniles shallow in intertidal zones, coastal protected boulder reefs, and often in small aggregations.

Top photo: A rare shot of a large group of *A. lineatus* swimming over a perfect reef top. Centre photo: A single Lined surgeonfish grazing algae.

Acanthurus lineatus **Palau**

Acanthurus lineatus **Sulawesi**

Orange-blotch surgeonfish

Le: To 35 cm. Di: Entire area. De: 2-50 m. Ge: A benthic feeder, grazing algae on rocks and sand among shallow reefs. Juveniles singly, often mixed with other juvenile surgeonfishes. Right: Adult. Below: Juvenile.

Acanthurus olivaceus **Singapore**

Acanthurus nigricans **Talaud I., Indonesia**

Velvet surgeonfish

Length: To 20 cm.
Distribution: Entire area.
Depth: 1-50 m.
General: Occurs mainly on outer reef crests in small to large aggregations when adult. Widespread in Indo-Pacific, showing no obvious differences throughout its large geographical range. Small juveniles are somewhat shy amongst large corals. Best distinguished from the similar *A. japonicus* (following species) by the relatively small white mark below the eye and sub-marginal yellow band in caudal fin.

Acanthurus japonicus **Okinawa, Japan**

White-nose surgeonfish

Le: To 14 cm. Di: Entire area. De: 3-20 m. Ge: In small to large aggregations on outer reef crests and slopes. The photo left shows *A. japonicus* in the foreground and the very similar *A. nigricans* in the background, both spp. occur in S-Japan together. Below: *A. japonicus* from the Philippines.

Acanthurus mata **Komodo I., Indonesia**

Elongate surgeonfish

Le: To 50 cm. Di: Entire area. De: 5-45 m. Ge: Occurs in groups along steep reef slopes often in turbid waters but also over sandy bottom in lagoons. Most of the time swimming in open water. Feeds primarily on zooplankton.

Pencilled surgeonfish

Le: To 50 cm. Di: Entire area. De: 3-100 m. Ge: Juveniles in large coastal estuaries, grazing on algae-covered rocks. Distinct from the following two similar species by white spine. Below: **Eye-spot surgeonfish** *A. bariene*, to 40 cm, entire area, 10-50+ m.

Acanthurus dussumieri **Palau**

Yellowfin surgeonfish

Le: To 56 cm. Di: Tropical Indo-Pacific. De: 5-90 m. Ge: One of the largest family members. Singly in greater depths. Feeds on diatom and detritus film on sand, filamentous algae, hydroids, pieces of fish. Juveniles coastal, on silt. Adults prefer clear coastal reefs. Below: Yellow pectoral fin is distinct.

Acanthurus xanthopterus **Bali, Indonesia**

Dark surgeonfish

Le: To 45 cm. Di: Entire area. De: 4-55 m. Ge: Black pectoral fin is distinct. Below: **Ringtail surgeonfish** *A. auranticavus,* to 35 cm, entire area 3-50 m. Indistinct, easily missed and mistaken for other similar species. Spot behind eye elongated, white band at tail base. Orange patch around tail spine.

Acanthurus blochii **Iriomote I., Japan**

Acanthurus maculiceps **Bali, Indonesia**

Spot-face surgeonfish

Le: To 35 cm. Di: Entire area. De: 1-20 m. Ge: Shallow, clear coastal reef flats and slopes. Usually in small groups or mixed with other species to feed on algae. Face spots distinct. Coloration varies from black to pale with fine lines. Below: **Horseshoe s.** *A. fowleri,* to 45 cm, Indonesia, Philippines, 6-50+ m.

Paracanthurus hepatus **Yap**

Palette surgeonfish

Le: To 30 cm.
Di: Malaysia to southern Japan, entire area.
De: 10-40 m.
Ge: Coastal to outer reef flats subject to currents and surge, often very shallow and usually in depths of less than 10 m. It occurs in loose aggregations 1 or 2 meters above the bottom and feeds on zooplankton. Juveniles and subadults secretive in *Acropora* thickets, swimming just above to feed on plankton, when alarmed, wedge themselves tightly among the branches. Large adults swim about openly and graze algae, sometimes forming schools to feed on plankton.

Paracanthurus hepatus **Bohol I., Philippines**

Yellow-tip bristletooth
Le: To 16 cm. Di: Entire area. De: 10-50 m. Ge: Clear coastal to outer reefs, on slopes and drop-offs with rich invertebrate growth. Juveniles singly, adults in small groups and mostly in 20+ m. Yellow tips on dorsal and anal fins distinct. Below: **Two-spot bristletooth** *C. binotatus*, to 20 cm, entire area, 2-50 m.

Ctenochaetus tominiensis **Flores, Indonesia**

Pacific sailfin tang
Le: To 40 cm. Di: Tropical west Pacific. De: 2-30 m. Ge: Juveniles in rich coastal coral and algae reefs and lagoons. Right: Adult. Below: Juvenile.

Zebrasoma veliferum **Nabucco I., Indonesia**

Brown tang
Le: To 20 cm. Di: Entire area. De: 0-15 m. Ge: Singly or in small groups in coral-rich areas of lagoons and seaward reefs. Juveniles secretively in rich coral growth. Below: This yellow colour variant is found in southern Japan, changes coloration there from brown to yellow.

Zebrasoma scopas **Sangihe I., Indonesia**

Naso annulatus **Okinawa, Japan**

White-margin unicornfish

Length: To 100 cm.
Distribution: Entire area.
Depth: 10-30+ m.
General: A shy species found on deep, clear coastal to outer reef slopes usually in small groups but occasionally in large schools. Large adults form large schools, they are identified by their long horn and the white margin on the caudal fin. Juveniles are grey with white caudal peduncle. Largest unicornfish, length to over 1 m, including the horn.

Naso lopezi **Flores, Indonesia**

Slender unicornfish

Length: To 60 cm.
Distribution: Widespread West Pacific, entire area.
Depth: 6-50+ m.
General: Always in small to moderate-sized schools along slopes or drop-offs subject to strong currents and leading to very deep water. Clear coastal to outer reef habitats, feeding well away from the substrate on plankton but sleep on the substrate by resting against rocks on sand. Adults are identified by their slender body shape and the numerous dusky spots on the upper half of the body.

Hump-nose unicornfish
Le: To 60 cm. Di: Widespread Indo-West Pacific. De: 2-45 m. Ge: Adults singly or in small groups on outer reefs, on crests near deep water or along steep walls. Males develop large hump in front of the eyes. Juveniles (below) and females plain greenish grey with numerous small dusky to yellow spots.

Naso tonganus **Talaud I., Indonesia**

Humpback unicornfish
Le: To 70 cm. Di: Entire area. De: 1-30 m. Ge: Occurs in pairs or in small groups in the shallow water area along the edges of steep outer reef slopes, especially in the shelter of large mushroom-shaped coral heads. Often mixed with other *Naso* spp. Right: School. Below: Male with horn (female without).

Naso brachycentron **Alor I., Indonesia**

Orangespine unicornfish
Le: To 50 cm. Di: Entire area. De: 2-90 m. Ge: In areas of coral, rock, rubble of lagoons and seaward reefs. It feeds primarily on leafy brown algae such as *Sargassum* and *Dictyota* spp. Right: School.
Below: **Little unicornfish**
N. minor, to 25 cm, entire area, 10-50 m.

Naso lituratus **Luzon, Philippines**

Naso hexacanthus **Banda Sea, Indonesia**

Sleek unicornfish

Le: To 75 cm. Di: Entire area. De: 10-137 m. Ge: Very common. In groups off outer reef drop-offs, feeds primarily on larger animals in the zooplankton. Not shy. While mating males are light-blue. Below: A single specimen.

Naso vlamingi **Sumatra**

Bignose unicornfish

Le: To 55 cm. Di: Entire area. De: 2-50 m. Ge: On seaward reefs, in deep lagoons. By day in mid-water off deep slopes, feeds on zooplankton. Not shy. While mating with blue stripes and spots, also at cleaning stations. Below: **Bluespine unicornfish** *N. unicornis,* 70 cm, entire area, 2-80 m.

Prionurus scalprus **Iriomote I., Japan**

Japanese sawtail

Le: To 40 cm. Di: Taiwan to S-Japan. De: 1-40 m. Ge: Coastal rocky-algal habitats, in small groups. Distinct: 4-5 peduncular plates of which 3-4 are black. Juveniles, females deep-bodied, with plain grey to nearly black body. Males (below) much more elongate, whitish ventrally with white streaks on head and body.

MOORISH IDOLS ZANCLIDAE

Moorish idol
Le: To 25 cm. Di: Entire area. De: 5-182 m. Ge: On hard substrates from turbid inner reef flats to clear seaward reefs. Usually in small groups but rarely in large schools of well over 100 animals. Feeds on sponges but also eats other animal and plant material. Right: School. Below: Single animal.

Zanclus cornutus **Palau**

RABBITFISHES SIGANIDAE

Gold-saddle rabbitfish
Le: To 40 cm. Di: Entire area. De: 3-35 m. Ge: Restricted to coastal rocky reefs with soft coral growth. Forms small groups, often seen hovering in one place in the shelter of reefs.

Rabbitfishes are primarily a tropical family, comprising 2 subgenera and about 30 species, broadly distributed in the Indo-Pacific. All species have pectoral fins with large, sharp, venomous spines. A sting is extremely painful and pain lasts for about a half hour. Applying heat brings relief.

Siganus guttatus **Sangihe I., Indonesia**

Java rabbitfish
Le: To 53 cm. Di: Entire area. De: 1-25 m. Ge: Often hard to recognise in the cloudy waters which it prefers. Has been observed feeding on jellyfishes. Below: **Maze rabbitfish** *S. vermiculatus,* to 50 cm, entire area, 1-45 m. Highly variable, from estuaries (juveniles) to deep clear coastal reefs.

Siganus javus **Derawan I., Indonesia**

Siganus punctatus **Bali, Indonesia**

Gold-spotted rabbitfish
Le: To 40 cm. Di: Entire area. De: 3-45 m. Ge: Juveniles on shallow reef flats, also in brackish water, adults pair in deeper coastal waters on mixed sand and rocky bottom. Below: **Peppered rabbitfish** *S. punctatissimus,* to 30 cm, entire area, 3-30 m. In pairs, appears plain from a distance.

Siganus puellus **Flores, Indonesia**

Masked rabbitfish
Le: To 30 cm. Di: Entire area. De: 3-35 m. Ge: Juveniles amongst branches of *Acropora* coral heads. Adults in pairs, usually secretive in staghorn corals in coral-rich lagoons. Below: **Double-barred rabbitfish** *S. virgatus,* to 30 cm, Indonesia, Philippines, 2-25 m. In coastal habitats.

Siganus vulpinus **Cebu, Philippines**

Fox-face
Le: To 25 cm. Di: Our area north to Philippines. De: 2-15 m. Ge: Unmistakable. Below: **Black-blotch fox-face** *S. unimaculatus,* to 25 cm, entire area, 3-45 m. Protected coastal waters. Juveniles shallow, in large schools, graze on algae. Adults may form pairs or can occur in schools over reefs.

Big-eye trevally

Le: To 85 cm. Di: Entire area. De: 3-100 m. Ge: Common, often in large schools. Small juveniles with dark crossbars. Large adults singly or in schools along deep drop-offs, usually deeper than 30 m. Juveniles in estuaries, forming schools in coastal bays as they grow and move out to deeper water. Below: **Black trevally** *C. lugubris,* to 80 cm, entire area, 5-50+ m. Black body marks distinct. Not common. On outer reef slopes and clear reefs.

The trevallies or jacks are a large tropical to temperate water family with about 25 genera and 140 species worldwide. Many are pelagic and wide-ranging, some enter brackish water.

Caranx sexfasciatus **Banda Sea, Indonesia**

Rainbow runner

Le: To 120 cm. Di: Entire area. De: 1-15 m. Ge: Pelagic, usually at or near the surface, over reefs or sometimes far offshore. Feeds on crustacean larvae and small pelagic fish. Small schools of juveniles may congregate in the vicinity of floating objects at sea. Adults are attracted by cleanerfish stations.
Below: **Golden pilot jack** *Gnathanodon speciosus,* to 110 cm, entire area, 2-50 m. Found inshore, including rocky reefs, deep lagoons and seaward reefs. Tiny juveniles live symbiotically among the tentacles of jellyfishes or swim below drifting clumps of algae. They often pilot sharks, rays or large groupers.

Elagatis bipinnulata **Nabucco I., Indonesia**

MACKERELS SCOMBRIDAE

Gymnosarda unicolor **Talaud I., Indonesia**

Dogtooth tuna
Le: To 180 cm. Di: Entire area. De: 3-100 m. Ge: Young often seen singly or a few together over the top of reef flats or in shallow coastal waters. Adults along outer reef drop-offs singly or in small schools, hunting other fishes, often taking reef planktivores which went a little too far out to feed.

The Scombridae (mackerels and tunas) comprise several subfamilies, 15 genera and about 50 species. Series of finlets behind the dorsal and anal fins, and 2 or more small keels on each side of the caudal peduncle are distinctive.

Rastrelliger kanagurta **Taiwan**

Mouth mackerel
Le: To 100 cm. Di: Entire area. De: 1-40 m. Ge: In schools, filters water for plankton, mouth wide open. Below: **Double-line mackerel** *Grammatorcynus bilineatus,* to 100 cm, entire area, 2-40 m. Singly, at drop-offs and seamounts, also in deep lagoons. Eats small pelagic fish and planktonic crustacea.

BARRACUDAS SPHYRAENIDAE

Sphyraena barracuda **Komodo I., Indonesia**

Great barracuda
Le: To 190 cm. Di: Entire area. De: 3-100 m. Ge: Juveniles occur among mangroves and in shallow sheltered inner reef areas. Adults are distinct by the dark blotches on their arrow-like body. This a solitary, diurnal species that may be quite curious and approach a diver closely but is no danger in clear water and under unprovoked conditions.

Barracudas are long, slender, silvery fishes with 2 dorsal fins clearly set apart, small scales and a pointed snout with large mouth and long, pointed teeth.

Yellowstripe barracuda
Le: To 50 cm. Di: Entire area. De: 2-45 m. Ge: Always in schools, extremely arrow-like body. Distinguishable from other species by a yellow lateral stripe. Below: **Yellowtail barracuda** *S. flavicauda,* to 45 cm, entire area, 5-70 m. Occurs in schools over lagoons and seaward reefs.

Sphyraena novaehollandiae Okinawa, Japan

Pickhandle barracuda
Le: To 150 cm. Di: Entire area. De: 10-40 m. Ge: In schools of a few specimens to hundreds. Favours areas of current, often near wrecks. With lateral bars. Below: Similar **Blackfin barracuda** *S. qenic,* to 130 cm, entire area, 10-90 m. In large tightly-packed schools, with about 20 lateral bars.

Sphyraena jello Palawan, Philippines

Sawtooth barracuda
Le: To 90 cm. Di: Area to Philippines, exact distribution not known. De: 0-15 m. Ge: In large groups, in current-swept lagoons, outer reefs. Below: **Forster's barracuda** *S. forsteri,* to 65 cm, entire area, 2-300 m. Solitary as adult, juveniles in schools. Black blotch on pectoral fin distinct.

Sphyraena putnamiae Sumatra

QUEST OF A DIVE-GUIDE

On Cabilao Island, a small island near Bohol in the Philippines, one can find a quiet diving resort, right next to the beach as there are many in this West Pacific country. Cabilao is famous for the schools of hammerhead sharks that roam these waters during particular periods of the year. The coral reef is not affected by El Niño and corals are blooming in each and every corner. On the reef and also at the walls of the drop-off a great number of huge gorgonians are waving in the current. Underwater photographer John Neuschwander reports on a fish that is only a little bit smaller than a hammerhead.

PHOTOS: JOHN NEUSCHWANDER

The members of a diving group examine a gorgonian in the waters off Cabilao for pygmy seahorses.

Bobet Cruz is a dive guide on Cabilao. He is an experienced scuba diver with an eye for the smallest details of the flora and fauna underwater. He had heard so much about pygmy seahorses and had seen so many pictures of them taken in Indonesian waters. And as Indonesia is not that far away from his home country, he started his quest for pygmy seahorses in the Philippines. Bobet was really convinced that they had to be there as well. He was searching his house reef daily, but he did not know in which gorgonian he had to look. So he looked for them in each and every gorgonian in all of his diving places month after month. Nobody could be more patient than him. It became a real obsession for Bobet. Only after one year he finally found his first 8-mm-long mini seahorse *Hippocampus bargibanti* in the gorgonian *Muricella plectana* on Christmas Eve 2000.

But his quest was not to come to an end. Where there is one, there must be more seahorses. And within a few weeks he found more and more of these tiny critters because now he knew where to look for them. In his diving area there are two big gorgonians with a vast number of seahorses, but they are so small, one can hardly see them. And they look like the polyps of the gorgonian. When the gorgonian bloomed and the polyps were out, I myself was not able to find the seahorses. But as soon as the polyps were retracted, I might have a chance to find the tiny fishes.

For an underwater photographer it is hard work to photograph them, it took me a lot of time and a lot of pictures. Every time you see one of those rosy 'fishlets' it turns its head away from you, showing only its back. Or it bends over in the wrong way between the branches of coral. It took me two dives and a lot of decompression (the gorgonians are at a depth of about 30 metres) to make one or two reasonable pictures. I only can advise all underwater photographers who are diving Malaysia, Viet Nam, and the region north to southern Japan to carefully check gorgonians. Indeed, pygmy seahorses are a challenge.

Above: The extraordinary Pygmy seahorse *Hippocampus bargibanti* is extremely well camouflaged in the gorgonian *Muricella plectana.* This one was the first specimen which Bobet discovered in the Philippines.

Right: There is reason to believe that Pygmy seahorses will not have the same fate as their larger relatives. As the photo from Hong Kong shows many seahorses and pipefishes end up in the shops of Chinese traders. In China these dried fishes are added to the meal like spices.

Left: Pygmy seahorses do not behave like actors but turn around when they see a camera. This photo is a good example for the camouflage of this tiny critter.

FLOUNDERS BOTHIDAE

Bothus pantherinus **Derawan I., Indonesia**

Leopard flounder

Le: To 30 cm. Di: Entire area. De: 1-100 m. Ge: Males develop long filaments in pectoral fin on ocular side, reaching caudal fin. Shallow coastal reef flats in sand patches ranging to very deep off-shore sand flats.

The Bothidae are a very large family with 15 genera and 90 species in the Indo-Pacific alone. Extremely compressed fishes, with eyes on left side of head. Ocular side pigmented to match surroundings, extremely well camouflaged, blind side unpigmented (underside). See further flounders in DISCOVERY IN SECRET BAY on p.65-66.

Bothus mancus **Singapore**

Peacock flounder

Le: To 45 cm. Di: Entire area. De: 1-85 m. Ge: Ocular pectoral fin large with greatly extended filamentous rays. Eyes far apart in large adults. Shallow coastal reefs and lagoons, on sand but often in rocky areas, to deep offshore on sand. Below: Larval stage.

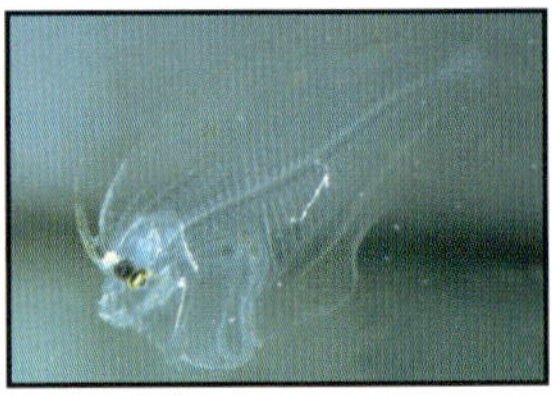

SOLES SOLEIDAE

Soleichthys heterorhinos **Alor I., Indonesia**

Banded sole

Le: To 15 cm. Di: Indonesia to S-Japan. De: 5-25 m. Ge: Rarely seen during the day, here shown swimming. In sheltered habitats, often silty areas. All soles have eyes on right side. Below: **Peacock sole** *Pardachirus pavoninus,* to 25 cm, entire area, 3-40 m.

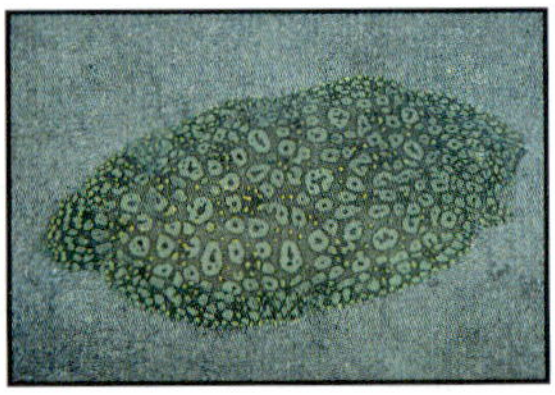

TRIGGERFISHES BALISTIDAE

The triggerfishes are a moderately large family, comprising 12 genera and about 40 species, occurring worldwide, but primarily in tropical Indo-Pacific. Large-headed fishes, first dorsal fin with 3 spines, of which 1st is stout and can be locked in upright position by 2nd spine, 3rd spine small. Pectoral fin usually small, paddle-like. Ventral fin rudimentary, encased with small scales, obvious, protruding from posterior end of ventral flap, the latter usually expandable to moderate extend. Jaws very strong, each with 8 long protruding close-set incisiform teeth. Most species are distinctly coloured and readily identified. They have pelagic larvae, a few are completely pelagic, some as juvenile only, and these are generally dull-looking, with little variation in growth. Benthic juveniles are distinctly coloured and usually differ greatly from adults. In most species there is virtually no difference between sexes. Males of some species are known to make nests and guard eggs vigorously, often attacking divers which venture into their territory. Many species live singly, found on or near reefs, however, nesting sites may consist of numerous males, each with a large territory. Only the pelagic-oceanic triggerfish *Canthidermis maculatus* schools. To escape danger, triggerfishes use narrow holes or ledges in which they wedge themselves head inwards, with the 1st dorsal spine ('trigger') locked in upright position, making it very difficult to be pulled out. Diet comprises diverse invertebrates, e.g. molluscs, echinoderms.

The **Filefishes** (p.237) are a large family of small to medium-sized fishes, comprising about 30 genera and an estimated 100 species, found in all major oceans. Their body is covered by tiny prickly scales, forming a tough leathery or velvet-like skin, and only in some species the scales are visible to the naked eye. A prominent and separate 1st dorsal fin spine is an obvious feature of most species, often very long, or armed with series of downward-directed barbs on edges, usually followed by a 2nd much smaller embedded spine and a narrow membrane. In nearly all species the spine can be locked in an upright position, and often fits into a shallow to deep groove when folded back. Relatively little is known about the biology of most species.

Titan triggerfish
Le: To 75 cm. Di: Entire area. De: 5-35 m. Ge: The largest triggerfish and one which can be very aggressive when guarding nests but may attack divers for no obvious reason. Best is to swim around large individuals at a distance, rather than passing close-by. Usually seen singly on coastal slopes adjacent to deep water. Below: Titan triggerfishes at a barbecue in an Indonesian village.

Balistoides viridescens **Gulf of Thailand**

Balistoides conspicillum **Ambon, Indonesia**

Clown triggerfish

Le: To 35 cm. Di: Entire area. De: 5-60 m. Ge: Readily identified by its large round white spots. Adults primarily along deep drop-offs, swimming openly about. Juveniles (below) secretive in caves, and usually deep.

Pseudobalistes fuscus **Luzon, Philippines**

Yellow-spotted triggerfish

Le: To 55 cm. Di: Entire area. De: 3-50 m. Ge: Deep reef outcrops, frequently over sand to feed on benthic invertebrates, blowing the sand to expose prey. Sometimes also called Blue triggerfish, relating to large adult males. Below: Juvenile.

Pseudobalistes flavomarginatus **Irian Jaya**

Yellow-margin triggerfish

Le: To 60 cm.
Di: Entire area. De: 2-50 m. Ge: Coastal reef slopes, adults singly, except when nesting on sand patches where they congregate but spread out with each in its own territory. Below: Juvenile.

Striped triggerfish
Le: To 30 cm. Di: Entire area. De: 6-50 m. Ge: Common in invertebrate-rich reef habitats with mixed coral and sponge growth, singly to small groups. Eats a great variety of invertebrates, including corals, echinoderms, and worms. Below: Male lacks stripes over top of snout. Right: Sleeping female at night.

Balistapus undulatus **Bali, Indonesia**

Redtooth triggerfish
Le: To 40 cm. Di: Entire area. De: 1-50 m. Ge: In enormous numbers on current-prone reef slopes and channels on hard bottoms. Rise high above reef to feed on plankton. Sleep in small holes with tail out, lobes crossed. Below: **Paddlefin triggerfish** *Melichthys vidua,* to 35 cm, entire area, 5-60 m.

Odonus niger **Sumatra**

Wedge-tail triggerfish
Le: To 25 cm. Di: Entire area. De: 1-20 m. Ge: Juvenile and adult similar, body pattern distinct. Common, singly, territorial, shallow, usually just below the intertidal. Juveniles in shallow reef flats or amongst small rock boulders. Below: **Blackblotch triggerfish** *R. verrucosus,* to 25 cm, our area, 3-45 m.

Rhinecanthus rectangulus **Mabul I., Malaysia**

Sufflamen chrysopterus Layang I., Malaysia

Half-moon triggerfish
Le: To 30 cm. Di: Entire area. De: 1-20 m. Ge: Adults on shallow reef flats, usually singly, but several may share a section of reef. Juveniles singly and shy, quickly diving into holes which are self-made by shifting sand. Below: **Boomerang triggerfish** *S. bursa,* to 30 cm, entire area, 3-90 m. Vertical 'V' distinct.

Xanthichthys caeruleolineatus Yonaguni I., Japan

Blue-lined triggerfish
Le: To 28 cm. Di: Southern part of area. De: 25-60 m. Ge: Blue line distinct. Male shown. Below: **Gilded triggerfish** *X. auromarginatus,* to 22 cm, all of area, 20-150 m. Singly or in pairs (shallow) or forming small to large aggregations in very deep zones. Males (below) with blue cheek and orange fin margins.

Xanthichthys mento Ogasawara Is., Japan

Crosshatch triggerfish
Le: To 29 cm. Di: S-Japan, subtropical oceanic islands. De: 40-90 m. Ge: In groups. Below: **Starry triggerfish** *Abalistes stellatus,* to 55 cm, entire area, 10-120 m. Body pattern identifies the species. Adults over sand staying loosely in touch with reefs. Juveniles in muddy habitats with reef outcrops.

Bristle-tail filefish

Length: To 7 cm.
Distribution: Entire area.
Depth: 1-25 m.
General: Found on coastal sand flats or slopes with small patches of weeds, large algae on rock, or among growths on jetty pylons like the pair on the photo at the right.

Acreichthys tomentosus Derawan I., Indonesia

Weedy filefish

Le: To 18 cm. Di: Entire area. De: 1-25 m. Ge: Skin flaps vary in density. Secretive in soft coral areas, extremely well camouflaged, often found accidentally. Right: Juvenile. Below: Adult.

Chaetoderma penicilligera Pulau Seribu, Indonesia

Long-nose filefish

Le: To 10 cm. Di: Entire area. De: 0.5-35 m. Ge: Coastal to outer reef lagoons, anywhere where large patches of *Acropora* corals are present. Adults pair whilst juveniles swim in small groups between or over the tips of the corals, feeding primarily on the polyps. Right: Juvenile. Below: Adult.

Oxymonacanthus longirostris Kerama I., Japan

Paraluteres prionurus **Sulawesi**

Mimic filefish
Le: To 10 cm. Di: Entire area. De: 1-25 m. Ge: Mimics *Canthigaster valentini* (p.240), a poisonous pufferfish in colour and shape in great detail, shares the same habitat. Below: **Ear-spot filefish** *Pervagor janthinosoma,* to 13 cm, entire area, 0.5-15 m. Shallow reef flats, rocky estuaries, secretive at rocks, sponges.

Pervagor melanocephalus **Singapore**

Black-head filefish
Le: To 10 cm. Di: Indonesia, Malaysia, Philippines. De: 10-45 m. Ge: Coastal to outer reefs, in rich invertebrate habitats with mixed soft corals and sponges, secretive in the shade. Usually in pairs, easily identified by orange body, dark head. Below: **White-line filefish** *P. nigrolineatus,* to 8 cm, 0.5-25 m.

Aluterus scriptus **Palau**

Scribbled filefish
Le: To 100 cm. Di: Entire area. De: 0.5-80 m. Ge: Small juveniles yellow with black spots, becoming ornamented with bright blue. Juveniles pelagic up to a large size, with floating weeds, in schools, adults singly, only rarely in small groups on coastal reefs. Below: A single individual at night.

Spotted boxfish
Le: To 20 cm. Di: Indo-Pacific, throughout our area. De: 3-35 m. Ge: Juvenile and female black with small white spots. Males derive from females through sex change (protogynous hermaphrodite). Right: Male. Below: Female. The bodies of boxfishes are covered with a toxic mucus.

Ostracion meleagris **Batangas, Philippines**

Striped boxfish
Le: To 12 cm. Di: Indonesia, Philippines. De: 3-40 m. Ge: Sheltered rich coral reef crests, slopes, caves along deep outer reef drop-offs. Juveniles with black stripes which break up with age and form a maze pattern on top and sides. Right: Male. Below: Female.

Ostracion solorensis **Ambon, Indonesia**

Yellow boxfish

Le: To 45 cm.
Di: Indo-Pacific, throughout our area. De: 1-45 m. Ge: Small juveniles (small photo below) cube-shaped, become moderately elongate with age. Caudal fin increasing in size with age and caudal peduncle becomes long and thick.

Ostracion cubicus **Bohol I., Philippines**

BOXFISHES — OSTRACIIDAE

Lactoria fornasini **Palau**

Thorny-back cowfish
Le: To 20 cm. Di: Entire area. De: 3-50 m. Ge: Green to brown with blue spots and scribbles. Rocky reefs and estuaries, usually singly on sand slopes with mixed sponge and weed habitats. Prominent spine in middle of back distinct. Below: **Longhorn cowfish** *L. cornuta*, 50 cm, entire area, 1-50 m.

PUFFERFISHES — TETRAODONTIDAE

Canthigaster valentini **Mabul I., Malaysia**

Saddled toby
Le: To 10 cm. Di: Entire area. De: 1-20 m. Ge: A distinct species, serves as model for non-poisonous mimics, e.g. the monacanthid *Paraluterus prionurus*. Common on shallow coastal to outer reefs, often in pairs, feeding on algae and a variety of small invertebrates. Males have blue lines radiating from the eye.

Family with about 20 genera and 100+ species. Common name pufferfishes applies to their ability to inflate themselves like balloons with water or air to deter predators.

Canthigaster coronata **Flores, Indonesia**

Crowned toby
Le: To 14 cm. Di: Entire area. De: 10-80 m. Ge: Similar to previous sp. but more colourful and usually in pairs on deep rubble slopes. Below: **Grey-top toby** *C. epilampra*, to 10 cm, entire area, 10-40 m, rare, deep, often 20+ m, singly on rich rubble slopes with sponges and soft corals.

False-eye toby
Le: To 10 cm. Di: Indonesia to Palau. De: 6-50 m. Ge: Along drop-offs in open caves but also on clear water coastal reef crests. Juveniles similar but more lined, and lines become more numerous and break up into spots with age. Replaced by **Solander's toby** *C. solandri* (small below) in some areas.

Canthigaster papua **Bali, Indonesia**

White-spotted puffer
Le: To 48 cm. Di: Entire area. De: 1-50 m. Ge: Reported to have a highly varied diet which includes Crown-of-thorns sea stars. Similar to the related *A. reticularis* but has spots, not lines, on snout. Usually in shallow water. Right: Unusual photo of congregating puffers to prepare for mass spawning.

Arothron hispidus **Bohol I., Philippines**

Manila puffer
Le: To 30 cm. Di: Entire area. De: 1-25 m. Ge: Adults singly or in pairs on coastal mud and sand flats and slopes, and in seagrass. Juveniles enter brackish water. Below: **Yellow-eye puffer** *A. immaculatus*, to 30 cm, entire area, 1-25 m. Coastal, found mainly in muddy areas and in the vicinity of seagrass beds.

Arothron manilensis **Sangihe I., Indonesia**

Black-spotted puffer
Le: To 30 cm. Di: Entire area. De: 3-25 m. Ge: The only sp. in the genus found mainly on coral reefs, often resting in barrel sponges. Highly variable, grey to brown or bright yellow to orange, spotted black to various degrees, and sometimes a mixture of colours. Occurs singly or in pairs. All puffers have the deadly poison tetrodotoxin in their organs and should not be eaten.

Arothron nigropunctatus **Bali, Indonesia**

Blue-spotted puffer
Le: To 80 cm. Di: Entire area. De: 5-40 m. Ge: Uncommon, easily confused with other large spp. This one is distinct by alternating brown and blue lines around eye with no radiating lines, and blue spots over body and fins. Found on gorgonian-rich coral reefs with currents. Below: **Starry puffer** *A. stellatus,* to 120 cm, 3-50 m. Bottom: **Map puffer** *A. mappa,* to 65 cm, 5-30 m.

Arothron caeruleopunctatus **Layang I., Malaysia**

Black-spotted porcupinefish
Le: To 71 cm. Di: Entire area. De: 5-100+ m. Ge: Spines erectile but rather short. Large individuals usually singly, often swimming high above the ground during the day, but usually near deep water. Juveniles associate with floating weeds where they may reach a moderate size before settling on the substrate.

The porcupinefishes are a small family comprising 6-7 genera and about 20 species. Like pufferfishes they are able to inflate themselves but have additional large spines which point outwards when inflated.

Diodon hystrix **Singapore**

Fine-spotted porcupinefish
Le: To 30 cm. Di: Entire area. De: 3-100 m. Ge: Spines long and narrow, erectile, many spines longer than eye-diameter, longest on forehead. Sheltered bays and estuaries, but also known from trawls. Below: **Blotched porcupinefish** *D. liturosus,* to 45 cm, 1-90 m.

Diodon holocanthus **Bali, Indonesia**

Rounded porcupinefish
Le: To 20 cm. Di: Entire area. De: 10-50+ m. Ge: Common but secretive, hiding deep in reefs during the day, but active at night. One of the smallest family members, eyes very large. Mainly on rich coastal reef slopes. Below: **Few-spine porcupinefish** *Chilomycterus reticularis*, to 55 cm, 0-100+ m.

Cyclichthys orbicularis **Luzon, Philippines**

TWO SHARP EYES WILL EARN YOU A COZY HOME

With more than 550 species worldwide, snapping shrimps form one of the most diversified shrimp families, the Alpheidae. Alpheids are very common in all tropical seas, including the coral reefs of the tropical western Pacific. Although very numerous most of these shrimps are difficult to encounter because they usually shelter in various cavities. The Crustacea specialist Arthur Anker reveals some of the secrets of symbioses between partner gobies and snapping shrimps.

RUDIE KUITER

Alpheus fenneri burrowing in muddy sand of deeper water is usually associated with the largest partner goby *Amblyeleotris fontanesii*. Flores, Indonesia.

Snapping shrimps of the genus *Alpheus* can be easily identified by the enlarged and asymmetrical claws, one of which is the famous snapping claw, and by the orbital hoods perfectly protecting the eyes. Many species make deep burrows in coral rubble, muddy or sandy substrates, usually on sandflats, seagrass meadow patches between the reefs or at the bases of coral. Gobies belong to the largest fish family, the Gobiidae, with more than 2,500 species worldwide. Most gobies are small and use different natural cavities or corals as shelters. Some make burrows or use abandoned burrows of other animals. Within this huge family of often very small fishes a small group composed of a dozen genera with about 100 species is well known for their mutualistic associations with burrowing snapping shrimps.

PHOTOS: TOMONORI HIRATA

Alpheus sp. with partner goby *Vanderhorstia mertensi*. Kochi, Japan.

Burrowing snapping shrimps usually use their snapping claw as a defensive weapon either against potential predators or against conspecific intruders of the same sex. While the claw is useful and quite effective against equally-sized or slightly larger opponents, it is of little help against larger predators such as carnivorous fishes, cephalopods or stomatopods. Contrary to the beliefs of most divers, alpheids are not completely blind, although it is true that the vision of some burrowing species is indeed reduced and the corneas almost lack pigmentation (that's why the

HELMUT DEBELIUS

This *Alpheus randalli* lives together with *Amblyeleotris yanoi*. Derawan, Indonesia.

Alpheus randalli with *Stonogobiops pentafasciata*. Kochi, Japan.

eyes sometimes appear whitish). For these reasons they would never leave their burrow during the day. This is also partly true for gobies, despite the fact that these lovely creatures have perfect vision and are able to swim very rapidly. Outside their shelter both gobies and shrimps would be extremely vulnerable. Thus we can imagine that the goby-shrimp association evolved mainly under the pressure of predation, and because these animals share the same habitats. This most parsimonious "cohabitation to coevolution" theory was proposed by the Japanese ecologist Harada. Gobies unable to make their own burrows plainly benefit from burrows made by *Alpheus* shrimps, but this is not all. The shrimps also clean their goby partners, as is shown in one of the photos. Furthermore, the goby profits indirectly

Alpheus sp. associated with two gobies *Vanderhorstia ornatissima,* showing the typical dumping behaviour. Okinawa, Japan.

Alpheus randalli with a pair of a spectacular, undescribed species of *Stonogobiops*. Kochi, Japan.

from the digging and dumping activity of the shrimp: many small sediment-dwelling crustaceans and worms become available as snacks to the fish.

Beside the obvious protection by warning, the shrimp's major advantage from an association with a goby is the optimal spacing and direction of burrow entrances, avoiding territorial conflicts with neighbours, especially in densely populated areas. Most people believe that the burrow entrances don't change much, but this is not true. If someone would insert a small bar in the sediment just behind the entrance in the morning, he would probably find the same bar several dozens of centimeters away from the entrance in the evening. This daily shift of the entrance can be explained by the restless burrowing activity of the shrimps, frequent collapses of the burrow's entrance and a new direction of the entrance dictated by the goby. After a week or so a burrow entrance "migrates" sometimes half a metre or more away from its original position!

Alpheus randalli associated with two individuals of the recently described partner goby *Tomiyamichthys alleni*. Kochi, Japan.

Another interesting detail ignored by most divers is that most burrows are occupied not by one but by two shrimps, each appearing from time to time at the burrow entrance. For a diver, however, it is almost impossible to know if the same shrimp appears or if two shrimps are alternating. If the shrimps often appear at the entrance and are equal in size, look closely to see on

Taxonomists have a lot of work to do: an undescribed *Alpheus* sp. associated with an undescribed *Flabelligobius* sp. Kochi, Japan.

which side the larger claw is situated: there is a fairly good chance that in the other individual it is on the other side! Rarely both shrimps - male and female - are seen outside the burrow.

The communication between gobies and shrimps is fascinating and complex, especially in so-called obligatory associations (when certain species of gobies and shrimps are always found together sharing the burrow). The gobies observe the surroundings at the entrance either posed on the bottom *(Amblyeleotris, Cryptocentrus, Ctenogobiops)* or mostly swimming above the entrance *(Lotilia, Stonogobiops)* (see photos). If a predator approaches too close to the entrance, the goby warns the shrimp by emitting a special warning signal with its tail, called "tail flick signal". The shrimp is able to perceive the signal through one of its antennal flagellae which is always in touch with the goby, especially when the shrimp is dumping sediment or foraging outside the burrow. This antennal contact can be seen in all photos on these pages. After the tail flick signal both partners usually disappear quickly into the burrow. Sometimes, when there is no danger in sight, the fish emits another signal, the so-called "guarantee-of-safety signal". While the tail flick is a sequence of rapid waves, the guarantee-of-safety signal is transmitted by a large amplitude tail movement. However, the simplest and most efficient signal is the so-called "head entry", a panic-like, very fast head-first retreat into the burrow. The goby first makes a looping and then dives into the burrow. Of course this movement warns the shrimps, though indirectly, but no less efficiently!

Usually the shrimps share their burrow with only one goby. Occasionally two gobies, probably a male and a female, inhabit the same burrow, as shown in one of the photos. A typical *Alpheus* burrow is a deep (0.5 -1 m), usually oblique tunnel, sometimes having horizontal extensions covering 1 square metre. Lateral tunnels may branch off the main tunnel, and there may be larger chambers or secondary issues. Often the tunnel penetrates into dead coral or under a huge coral block where the major portion of the burrow is situated. It is impossible to excavate such a tunnel, as experienced by many people including the author of these pages! During the night the burrow entrance is closed with sediment and both shrimp and goby stay in the burrow until the next morning. Thus, the alpheid burrow is one of the safest sheltering places in a coral reef. The excavation of a burrow begins just after the settlement of the shrimp, in other words soon after the metamorphosis from the ultimate planktonic larval stage to the small crawling juvenile. It has been shown that at least in some species the first mate is acquired soon after the beginning of burrowing activity - in this way building the burrow continues much faster. It is possible that in absence of a disturbance from the outside (predation, conflict with a dominant neighbour etc.) these alpheids remain monogamous, keeping a life-long mating partnership with the first partner encountered. In contrast to this, the gobies often leave their shrimp partners and are replaced by others. Competition for burrows among gobies is strong and therefore the recruitment of new gobies is quite fast (1-2 weeks).

While the ecological and behavioural aspects of these associations were studied in great detail, the taxonomy of goby-associated shrimps unfortuntaely was almost completely neglected, resulting in erroneous or incomplete identifications, invalid names and other confusions. More than a dozen species await scientific description, and there are probably more! Like in many other shrimps the colour pattern seems to be species-specific and many species can be easily recognised by their unique and hence diagnostic colour patterns (see photos). Also, we still have very scarce data on the distribution of most species and their specific partners, which is one of the most interesting aspects to study.

CRUSTACEANS CRUSTACEA

Crustaceans are one of the truly dominant animal groups in coral reefs. Due to the tiny size and cryptic habits of many species, however, it is easy to underestimate their impact on the reef's ecology. The majority of crustaceans encountered in coral reefs are decapods. Decapod literally means 'ten legs'. These are four pairs of walking legs and the first pair which is often strongly developed and in many species has terminal claws. Such clawed legs are called chelipeds. Shrimps (often called prawns), lobsters, and crabs are all well-known members of this group. For the most part of their life they are secretive creatures that remain hidden for long periods in burrows, crevices, under slabs of dead coral, or inside empty mollusc shells. They form an integral part of the food chain, being hunted by larger predators, especially fishes. They have evolved a largely inconspicuous life style in response to this predation pressure. Generally the best time to observe crustaceans is at night, when many species emerge from their retreats to feed. The largest of the about 1,000 decapod species known to date has a carapace length of up to 50 cm and its legs span up to 3 metres. This is a cold water species, but size is not everything. Observing and photographing the multitude of colourful crustaceans in tropical reef environments will always yield new insights into the complex interactions of crustaceans with their hosts, friends, and foes. Anyone, who wants to know more about crustaceans, should have a look at the CRUSTACEA GUIDE OF THE WORLD by the same author.

BOXER SHRIMPS STENOPODIDAE

Stenopus hispidus **Bunaken I., Indonesia**

Banded boxer shrimp
Le: To 4 cm. Di: Entire area, throughout tropics in all three oceans. Ge: First three pairs of walking legs clawed, third pair with especially large claws, always held spread out, hence the name boxer shrimps. Body and legs characteristically banded white and brown. Always in pairs. In reefs from 6 m down, perching at the opening of a cave or large sponge, waving long, thin, white antennae to attract fish (or uw-photographers) which then are cleaned from parasites and old skin. The photo left shows an egg-carrying female in the background, and a male in the foreground. Below: Cleaning a moray.

BOXER SHRIMPS STENOPODIDAE

Blue-legged boxer shrimp

Le: To 3 cm. Di: Entire area, tropical West Pacific. Ge: Singly or in pairs on reefs, hiding in caves. Without externally visible sexual dimorphism. Very shy. Below: *S. tenuirostris,* to 2 cm, trop. IP, usually in pairs, one of the smallest genus members.

Stenopus cyanoscelis **Ambon, Indonesia**

CLEANER SHRIMPS HIPPOLYTIDAE

Yellow-legged cleaner shrimp

Le: To 1.5 cm. Di: Indonesia to Japan and Philippines. Ge: From shallow water to at least 62 m. Genus monotypic, species uncommon. Below: *Lysmata vittata,* to 3.5 cm, entire area, littoral to 54 m.

Lysmatella prima **Sulawesi**

White-banded cleaner shrimp

Le: To 6 cm. Di: Entire area, widely distributed in Indo-Pacific. Ge: One of the most important invertebrate cleaners of reef fishes. Often in pairs, also in groups of up to 100. Family members are known to clean especially large, stationary reef fishes (groupers, morays).

Lysmata amboinensis **Palau**

Lysmata debelius **Bali, Indonesia**

Scarlet cleaner shrimp

Le: To 5 cm. Di: Entire area, Indian Ocean to Central Pacific. Ge: Usually found below 10 m. In pairs or small groups, defends territories against conspecifics. Without externally visible sexual dimorphism. Although not the most important invertebrate cleaner, readily cleans fishes.

Parhippolyte uveae **Derawan I., Indonesia**

Cave shrimp

Le: To 4 cm. Di: Entire area, widespread in Indo-Pacific. Ge: Tail fan without white markings, often misidentified as the closely related *P. mistica* (below, tail fan with white markings). Found in caves from 3-30 m which it leaves only at night. Both species are cleaners.

Thor amboinensis **Bali, Indonesia**

Squat shrimp

Le: To 2 cm. Di: Entire area to Central Pacific. Ge: The common squat shrimp is a commensal on corals, anemones, e.g. *Heteractis.* Abdomen held up almost vertically. Female almost twice the size of male. Lives in pairs or small mixed groups on host. This is not a cleaning species.

CLEANER SHRIMPS HIPPOLYTIDAE

Saw blade shrimp

Le: To 5 cm.
Di: Entire area, Indonesia to Japan and New Caledonia.
Ge: On black coral, hard to spot. This shrimp is extremely elongated, its body is transparent with irregular transverse camouflaging bands. The rostrum makes up for about a third of the total length but is less than twice as long as the remainder of the carapace. Rostrum without dorsal teeth but with 10-30 ventral teeth, giving it the appearance of a tiny saw blade. The abdominal somites nos. 3-5 each carry a dorsal tooth.

Tozeuma armatum **Sulawesi**

Common marble shrimp

Le: To 4 cm. Di: Entire area, widely distributed in Indo-Pacific. Ge: Shown here is the typical coloration of the nominal species of the very variable *S. marmoratus*-group. Large photo at the right shows a male, the small photo below shows a female.

Saron marmoratus **Derawan I., Indonesia**

SNAPPING SHRIMPS ALPHEIDAE

Soror snapping shrimp
Le: To 3.5 cm. Di: Philippines, probably widespread in Indo-West Pacific. Ge: Walking legs purple, large ocellus. Below: Right one is large snapping claw.

Alpheus soror **Papua New Guinea**

Alpheus ochrostriatus **Cebu, Philippines**

Fine-striped snapping shrimp
Le: To 5 cm. Di: Philippines, Papua New Guinea, widespread in Indo-West Pacific. Ge: Snapping shrimps live associated with various gobies in burrows dug out by one or a pair of the shrimp and guarded by the keen-eyed goby (singly or in pairs). The shrimp has much less acute vision and always keeps contact to its fish partner with one antenna in order to get warnings of possible threats as early as possible. The most common partner shrimp is *A. randalli,* shown in TWO SHARP EYES WILL EARN YOU A COZY HOME on p.244-247.

Alpheus djeddensis **Palau**

Djeddah snapping shrimp
Le: To 5 cm. Di: Entire area. Ge: Found shallow in coral reefs. Like previous species with two white saddles, a black blotch on each side of the carapace is distinctive. Lives together with several goby species, e.g. *Amblyeleotris steinitzi.* Below: *A. bellulus,* to 5 cm, Indonesia to Japan.

Alpheus rubromaculatus **Mabul I., Malaysia**

Red spot snapping shrimp
Le: To 4 cm. Di: Entire area, Red Sea to Tuvalu. Ge: Easily identified by red to purple spots all over body and legs. Almost exclusively lives together with the goby *Lotilia graciliosa.* Below: *A. djiboutensis,* to 4 cm, widespread in Indo-Pacific, coloration distinct.

HINGE-BEAK SHRIMPS RHYNCHOCINETIDAE

Durban hinge-beak shrimp

Le: To 4 cm. Di: Entire area, north to Japan. Ge: Common, in crevices. Incorrectly reported as *R. uritai,* differs in having a very long, toothed rostrum and white inner lines as wide as the red ones. Below: *R. conspiciocellus,* to 4.5 cm, warm temperate Japan, photo Okinawa.

Rhynchocinetes durbanensis **Bohol I., Philippines**

Striped hinge-beak shrimp

Le: To 6.5 cm. Di: New Caledonia to Philippines and Ryukyu Islands, S-Japan. Ge: On rocky reefs in 1-20 m. Rostrum hinged, moveable (hence family name). Below: *C. reticulatus,* to 6.5 cm, entire area, widespread in Indo-Pacific, most often seen on vertical walls near caves.

Cinetorhynchus striatus **Bohol I., Philippines**

COMMENSAL SHRIMPS PALAEMONIDAE

Plumed rock shrimp

Le: To 3 cm. Di: Entire area, Indonesia and New Caledonia to S-Japan, widespread in Indo-West Pacific. Ge: Slender, pattern distinct. Tufts of cirri on antennal scales and abdomen. Shallow to 15 m, on attached plants in reefs.

Palaemonids are a speciose family divided into two subfamilies, Rock shrimps (Palaemoninae, mainly cold water) and Commensal shrimps (Pontoniinae, mainly tropical). The latter are commensals living in non-parasitic community with other animals, often anemones.

Leander plumosus **Flores, Indonesia**

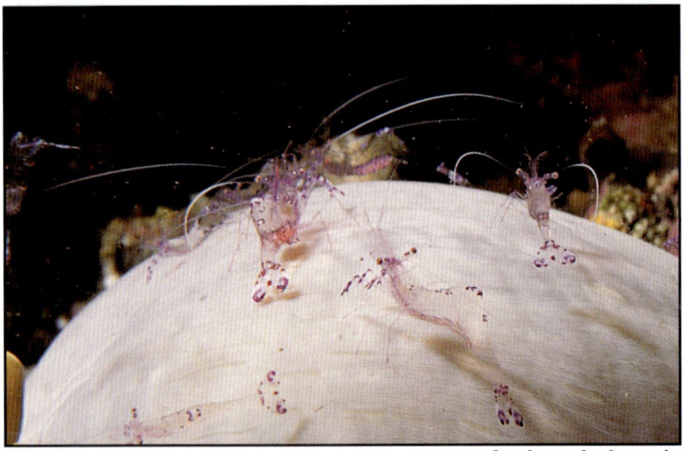

Periclimenes tosaensis **Ambon, Indonesia**

Tosa commensal shrimp
Le: To 2.5 cm. Di: Entire area, Indonesia to Japan. Ge: Often referred to as *P. holthuisi* but much more common and with different colour pattern. In up to large groups on coelenterate hosts, mainly anemones, also on bubble corals *(Plerogyra* spp.), mushroom corals *(Heliofungia* spp.), *Cassiopeia* jellies.

Periclimenes venustus **Batangas, Philippines**

Pepper-and-salt commensal shrimp
Le: To 2.5 cm. Di: Entire area, Indonesia to Philippines and Japan. Ge: Transparent, with numerous small white and blue spots. Colour pattern probably varies with host. Associated with anemones and the mushroom coral *Heliofungia actiniformis*. Below: Colour variant.

Periclimenes commensalis **Sulawesi**

Crinoid commensal shrimp
Le: To 1.8 cm. Di: Entire area, Indonesia to tropical Japan and oceanic islands. Ge: Commensal of comatulid crinoids (feather stars), protected by cryptic coloration, host active only at night, with only few predators. Below: *P. cornutus,* to 1.5 cm, entire area, living on crinoids.

Short hand commensal shrimp
Le: To 4 cm. Di: Indonesia and South China Sea to oceanic islands. Ge: Transparent, with large white blotches (largest on head), and 5 dark and orange ocelli on tail fan. With anemones, especially *Cryptodendrum adhaesivum*. Below: *P. kororensis,* to 4 cm, living with hydroids.

Periclimenes brevicarpalis **Gulf of Thailand**

Imperator commensal shrimp
Le: To 2 cm. Di: Entire area, widely distributed in Indo-Pacific. Ge: Living with a variety of invertebrate hosts, mainly and often in pairs on the colourful nudibranch *Hexabranchus sanguineus* (Spanish dancer). Below: *P. lanipes,* to 2.5 cm, southern part of area.

Periclimenes imperator **Sipadan I., Malaysia**

Coral commensal shrimp
Le: To 2.5 cm. Di: Indonesia, Hong Kong, Japan. Ge: Found associated with *Heliofungia* and *Goniopora* corals. Genus has only two species (the other from Australia). Below: *Pliopontonia furtiva,* to 1.8 cm, Indonesia, Philippines, the sole pontoniine living with disc anemones.

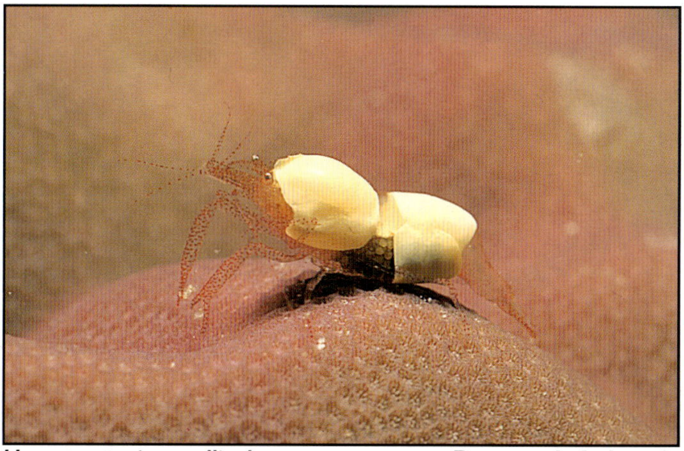

Hamopontonia corallicola **Derawan I., Indonesia**

URCHIN HIKERS

What use are the most beautiful spines and armour plates to echinoderms if the louse already thrives in the pelt? Such is the fate of the widespread Indo-Pacific **Fire urchin** *Asthenosoma varium* of the family Echinothuriidae. Despite its poisonous spines there are a variety of invertebrates - mainly crustaceans - and fishes that live on or close to the surface of its test. Most of them obviously do not harm their host or inflict only minor damage; others, however, such as the eulimid gastropod *Luetzenia asthenosomae*, are true parasites. Jim "Sherlock" Black has discovered all of them around Komodo.

PHOTOS: JIM BLACK

Above: *Periclimenes colemani* (family Palaemonidae) is usually found in pairs exclusively on the Fire urchin where it occupies a patch which it has cleared of tube feet and spines. The shrimps move amongst the poisonous spines and pedicellaria without inculting any harm. The larger shrimp is the female; its carapace is awkwardly swollen due to a parasitic bopyrid isopod which has infested its gills. The branchial parasite also lives in pairs, and sometimes even two pairs can be found on a single shrimp (male or female), one on each side of its carapace. The female isopod is a large, asymmetrically segmented animal filling the shrimp's gill chamber. The male is a tiny, symmetrical animal whose sole function in life is to fertilise the eggs produced by the female.

Left: The white and dark brown striped **Urchin crab** *Zebrida adamsii* (family Eumedonidae) is another "guest" of the Fire urchin. In contrast to *P. colemani* it lives in association with several sea urchin species. The last segment of the leg forms a hook to hold onto the spines of the sea urchin. It is found singly or in pairs and seems not to harm its host.

Right: The **Urchin cardinalfish** *Siphamia unicolor* (family Apogonidae) is a vertebrate commensal of the Fire urchin. It lives in small groups under the protection of the sea urchin's spines. Whether the Fire urchin has any advantage from the fishes' presence is not known. Other members of the genus *Siphamia* are found exclusively among the spines of the black *Diadema setosum* sea urchins or on the spiny **Crown-of-Thorns** sea star *Acanthaster planci*.

Left: The radula (rasping tongue) of the snail *Luetzenia asthenosomae* (family Eulimidae) is reduced in favour of a protrusible trunk as a consequence of its parasitic lifestyle. The gastropod (shown with its eggs) is an obligate parasite on different species of sea urchins of the genus *Asthenosoma*. It feeds on the urchin's ambulacral feet and other external soft tissues. Numerous species of eulimids have specialised as parasites of various echinoderms. All have conservatively built colourless shells.

Right: The commensal shrimp *Allopontonia iaini* (family Palaemonidae) is found on the Indo-Pacific Fire urchin and the West Pacific sea urchin *Salmacis belli*. Commensalism is a form of living together which clearly has advantages for one of the partners (the shrimp is protected by the spines of the urchin). The other species involved has no (known) advantage, but no disadvantage either. After some research commensal relationships often turn out to be true symbioses.

BUMBLEBEE SHRIMPS GNATHOPHYLLIDAE

Phyllognathia ceratophthalmus **Gulf of Thailand**

Horned bumblebee shrimp
Le: To 2 cm. Di: Entire area, Indonesia to Japan. Ge: Colour pattern of family members distinct. Agile and quick to hide, rarely seen. Found in 2-20 m. Below: *Gnathophyllum americanum,* to 3 cm, cosmopolitan in tropical seas, down to 50 m, occasionally with echinoderms.

HARLEQUIN SHRIMPS HYMENOCERIDAE

Hymenocera elegans **Singapore**

Harlequin shrimp
Le: To 5 cm. Di: Indonesia, widely distributed from Red Sea to N-Australia. Ge: Shape and coloration unmistakable, in intertidal and subtidal of coral reefs. Always in pairs, which are kept together by the female's pheromones. Pairs obviously territorial, may stay together for years. Large chelipeds (claws) used for communicative displays. The pair make a courtship dance prior to mating. Females are larger than males, and differ slightly in the second colour patch on side of abdomen.

REEF LOBSTERS ENOPLOMETIDAE

Enoplometopus debelius **Bali, Indonesia**

Debelius' reef lobster
Le: To 10 cm. Di: Entire area, West and Central Pacific. Ge: Body covered with purple spots. Common in Indonesia, Philippines. Below: *E. occidentalis,* to 12 cm, entire area. Reef lobsters hide in caves and crevices by day to come out to forage at night.

SPINY LOBSTERS — PALINURIDAE

Scalloped spiny lobster
Le: To 33 cm. Di: Entire area, Indonesia to New Caledonia and Japan. Ge: Found in 1-90, mainly 1-5 m among rocks, often in the surf zone. Like most family members nocturnal and gregarious. Below: *P. japonicus,* to 30 cm, Taiwan and East China Sea to Japan and Korea, in 1-15 m on rocky bottoms.

Panulirus homarus **Bali, Indonesia**

Ornate spiny lobster
Le: To 50 cm. Di: Entire area to S-Japan. Ge: One of the largest genus members. Found in 1-50, mainly 1-8 m on sand, mud, also rocky bottoms and coral reefs, often near estuaries in shallow, sometimes turbid waters. Below: *P. polyphagus,* to 40 cm, southern part of area north to Philippines, 3-90 m.

Panulirus ornatus **Ambon, Indonesia**

SLIPPER LOBSTERS — SCYLLARIDAE

Sculptured slipper lobster
Le: To 23 cm. Di: Entire area north to Philippines and Japan. Ge: Family members with flat, rounded antennae, used to dig up hard-shelled molluscs. Below: *P. japonicus,* to 16 cm, Ryukyu Is. to NW-Japan, in reefs.

Parribacus antarcticus **Bunaken I., Indonesia**

HERMIT CRABS — DIOGENIDAE

Aniculus aniculus **Alor I., Indonesia**

Bristly hermit crab
Le: To 10 cm. Di: Entire area, West and Central Pacific. Ge: All genus members are large and very colourful. Legs appear scaly, rim of scales with rows of long and light-coloured bristles. Not common, active during the day. Hermits use empty gastropod shells as permanent shelter. Below: *Calcinus elegans,* to 1.8 cm, entire area, intertidal.

Dardanus pedunculatus **Cebu, Philippines**

Anemone hermit crab
Le: To 5 cm. Di: Entire area, Indo-West Pacific. Ge: Very common, eye stalks striped red and white, eyes green. Shell covered with anemones that give protection with stinging tentacles in return for being moved to food sources (symbiosis). Below: *D. lagopodes,* 3 cm, entire area, eye stalks yellow.

SQUAT LOBSTERS — GALATHEIDAE

Lauriea siagiani **Bali, Indonesia**

Pink squat lobster
Le: To 1.4 cm. Di: Indonesia, Philippines. Ge: In sponges *(Xestospongia* sp., photo) Squat lobsters are densely covered with fine white or pink bristles. Related to hermits rather than true (spiny) lobsters. Below: *Galathea* sp., to 1.5 cm, Indonesia, yet unnamed.

PORCELAIN CRABS — PORCELLANIDAE

Oshima's porcelain crab
Le: To 2.5 cm. Di: Entire area, Indo-West Pacific. Ge: With large spots. Almost always in pairs on sea anmeones. Colour pattern varies from place to place. Below: *N. maculatus,* to 3 cm, area north to Philippines, with small spots. Among tentacles or under rim of anemones *(Cryptodendrum, Entacmaea).*

Neopetrolisthes oshimai — Derawan I., Indonesia

ROUND CRABS — XANTHIDAE

Red round crab
Width: To 3 cm. Di: Entire area, Indonesia to Philippines and Japan. Ge: This bright red round crab is found in rubble areas down to a depth of about 10 m. The lateral spines on the smooth carapace are low, giving it an angular appearance. The rare photo at the right shows a female of this species releasing young into the water. Before that, the crab has been guarding its eggs underneath the body for an extended period of time. Small photo immediately below: *Carpilius convexus,* to 9 cm, entire area, carapace and legs smooth except for orange peel texture. Bottom small photo: *C. maculatus,* to 9 cm, entire area, carapace smooth, four blunt spines between eyes.

Neoliomera insularis — Lembeh Strait, Indonesia

CORAL CRABS — TRAPEZIIDAE

Trapezia lutea **Bali, Indonesia**

Orange coral crab

Width: To 2 cm. Di: West Pacific. Ge: Carapace round, convex, eyes can be retracted into pits. First pair of legs enlarged, claws heavy with bristly outer surface. Common, by day in pairs, also with juveniles on branches of *Stylophora* and *Pocillopora* corals. Below: *T. rufopunctata,* to 1.5 cm, with about 200 pink dots.

SPIDER CRABS — MAJIDAE

Xenocarcinus conicus **Ambon, Indonesia**

Conical spider crab

Le: To 1.5 cm. Di: Entire area, Indonesia to Philippines and Japan. Ge: Long, triangular, variable grey to red, back often with long white stripe. On gorgonians, black coral, hydroids. Below: *X. tuberculatus,* to 1 cm, carapace with large spines.

SWIMMING CRABS — PORTUNIDAE

Charybdis acutifrons **Bohol I., Philippines**

Straight-faced swimming crab

Width: To 6 cm. Di: Malaysia and Indonesia to Japan. Ge: Found in a depth of 10-30 m on rocky bottoms and also in coral reefs. Coloration dark olive green with dark red blotches on sides of carapace, six sharp spines between eyes.

Portunids are able to swim with the paddle-like flattened distal segments on their hindmost pair of legs. They are swift and agile hunters, attacking even fast-moving prey such as bony fishes after resting motionlessly on the ground.

SMASHING MANTIS SHRIMPS — ODONTODACTYLIDAE

Purple-blotched smasher
Le: To 8 cm. Di: West Pacific, common in Indonesia and Malaysia. Ge: Colour pattern of outer branch (exopod) of tail fan distinct, dark black patch followed by purplish-pink terminal segment. Probably the best-known smashing mantis shrimp is the very colourful, widely distributed (E-Africa to Guam) *O. scyllarus* with its unique coloration of red, blue, and green.

There are two main groups of mantis shrimp according to the shape and function of the claws: Smashers have heavy, strongly calcified claws and prey on hard-shelled organisms (mussels, crustaceans); spearers have long, slender claws with spikes to catch slippery prey (fishes).

Odontodactylus latirostris — **Bali, Indonesia**

SPEARING MANTIS SHRIMPS — LYSIOSQUILLIDAE

Orange spearer
Le: To 25 cm. Di: Indonesia, Philippines. Ge: Only recently described. Mantis shrimps live in self-dug burrows and hunt from the entrance or in the vicinity. Small photo immediately below: *L. lisa* with the anemone shrimp *Thor amboinensis* (p.250) acting as a cleaner. Bottom small photo: *L. maculata*, to 38 cm, entire area, in pairs in u-shaped burrows.

Lysiosquillina lisa — **Palau**

EATEN BEAUTY

Giant clams are very popular among reef aquarists, probably because of their beautiful colours, their shape and their quiet presence, among other factors. For many divers and snorkellers, tridacnid clams are among the most fascinating organisms in the reef. Unfortunately, divers and reef aquarists are not the only people who admire tridacnid clams. In many tropical countries all over the world, giant clams are found in the fish markets among tilapia, squid, shrimps and crabs. For many aquarists it might be hard to believe that in other countries their highly desirable reef tank "jewels" end up in the cooking pot. But to many people giant clams are nothing more than larger varieties that feed the whole family. Journalist and marine aquarist Daniel Knop reports.

PHOTOS: DANIEL KNOP

This special type of giant clam aquarium is situated in a restaurant in Cebu (Philippines). Guests can chose a clam for their own consumption.

Consumption of giant clams by coastal human populations is not, however, the biggest threat to the natural stocks. These clams have long been eaten by native populations in tropical countries. These people often depend on seafoods to a great extent and they can simply not afford to leave the clams in the sea. They only collect what they need at the moment, because the available technology for refrigerating their catch is very limited. So, normally they collect relatively small numbers of clams.

The natives usually make use of the whole clam and eat everything except the kidneys (we have been told that clam kidneys have a very bitter taste). The shells are used mainly as a bowl or pot, as a baby bath tub or as a water trough for pigs. But mostly, clams are not regarded as a delicacy in these countries, and people eat tridacnid clams whenever they have no fish, or there is no money to buy fish. Normally, 1 kg of clam meat is much cheaper than the equivalent weight of fish.

The consumption of clams by native populations is not among the biggest threats to natural clam populations because those people use very primitive methods to harvest the clams. This limits their catch and also the areas where they can collect. Admittedly, the situation is worse in some countries where the population is very dense. For example, in some Philippine coastal areas, clams are extinct, and three species are even extinct around the entire country. The number of remaining giant clam wild stocks normally decreases the more closely a habitat is located to the shoreline, making the clams conveniently accessible to the coastal population. Near the coast, the only giant clams that can be found are normally well-hidden specimens (e.g., clams that are located under a table of *Acropora hyacinthus* stony coral and therefore hard to detect). Farther from the coast, out of the reach of coastal citizens, there is a larger number of remaining giant clam wild stock that can propagate and, in some instances, even attempt to re-populate the coastal zone with baby clams (which, of course, have a very limited chance to reach adulthood).

The commercial collection of giant clams is by far a bigger threat to natural populations. In these instances, huge, ocean-going ships are generally used, so even very remote clam habitats are accessible. Commercial fishermen use specialised collecting techniques and tools that are able to maximise their catch and open the clams quickly. Many use

Giant clam meat is cheaper than fish. Just 4 US $ were asked for these clams.

special knifes with long handles that enable the fishermen to kill the giant clams very effectively. This knife can pass through the small gap between the upper shell valve margins that exists even when *T. gigas* is closed. A huge specimen of this giant clam species can opened quickly.

These giant clam harvesting vessels are normally stationed in the Far East, mostly Taiwan, where the adductor muscle of a giant clam is regarded as a delicacy. People even believe in its aphrodisiac effects and pay huge sums for it. These people are not interested in the other parts of the clams like the native fishermen are. The siphonal mantle is quite tough and not in great demand. Therefore, most of the commercial giant clam collectors concentrate on the adductor muscle and leave the rest of the clam behind. The adductor muscle can easily be frozen and sold as fish meat. Bringing the entire clam on board the vessel would be very time-consuming and take up a lot of space, and would certainly limit the number of clams that could be harvested. However, when these commercial fishermen restrict their harvest to the adductor muscle only, a clam harvesting vessel can easily "harvest" 10,000 to 30,000 clams, depending on the size of the ship. All the "leftovers" of the opened giant clams will remain on the floor of the sea, providing food for the fish.

These ships normally operate unregulated. It is simply impossible to patrol all natural giant clam habitats. Also, the countries around which wild stocks of giant clams are normally found are relatively poor and therefore cannot invest much money in conservation enforcement. Furthermore, it is mostly the larger specimens that are being killed, because they are easily detected and yield a larger harvest. This is a great threat for the wild clam colonies because it limits their ability to reproduce. The remaining smaller clams have not yet reached their female maturity and will probably be harvested long before they can reproduce and preserve the species for the future. The pity is that if adult specimens were left in place, they could be constantly producing offspring within the colonised area and even in some remote locations via dispersal of clam larvae.

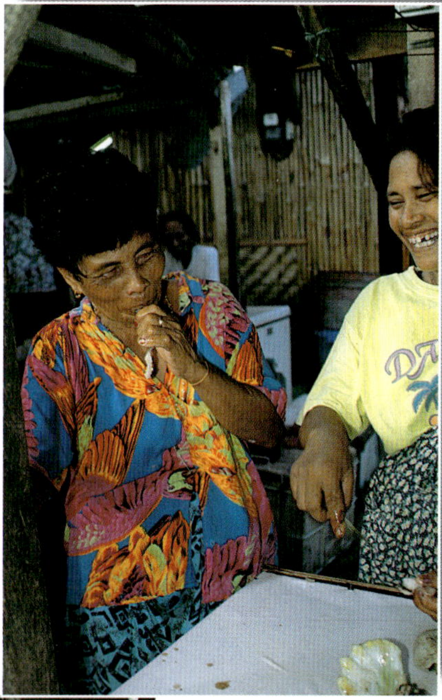

Giant clams as "fast food" in a Philippine fish market.

This special type of a "fence" seen in the Philippines shows how large amounts of giant clams are being eaten in tropical countries.

This threat has led to efforts to develop breeding methods and facilities for artificial giant clam culture. It has not been the primary goal of those efforts to supply the aquarium market with farm-raised clams. The main goal is to re-seed depleted natural clam habitats and even initiate new populations of certain giant clam species around countries where they are totally extinct. In order to try and prevent newly re-seeded clam populations from being harvested by Taiwanese clam collecting vessels, these facilities plan to supply the food industry with farm-raised clams as well, and the aquarium market is provided with clams to create a cash backflow and support the farming projects.

Giant clams have to suffer for silly decorative items sold by the curio trade.

MOLLUSCS MOLLUSCA

Molluscs are characterised by a muscular mantle (the outer part of the soft body), a radula (a tooth-covered rasping tongue working much like a conveyor belt), and a muscular foot that originally serves in locomotion but may be modified for diverse uses. Molluscs are adapted to live almost anywhere in barren, lifeless zones, where food is non-existent or inaccessible. Thus, sandy areas are heavens for molluscs with long siphons and bodies adapted for burrowing, such as a majority of bivalves, and many gastropods. A mixture of sand and mud will provide an even better habitat for burrowers because it contains more food. This is the ideal environment for numerous bivalve groups, particularly those with delicately constructed shells. Molluscs with stronger shells are associated with rocky shores and their rough conditions. The phylum Mollusca comprises several classes, covered here are bivalves (Bivalvia or Lamellibranchiata), snails and slugs (Gastropoda), and cephalopods like octopus, cuttlefish, and squid (Cephalopoda).

Bivalves are found from the tidal zone down. All have two shells (left and right), connected by a group-specific type of hinge and flexible bands (ligaments). Most are sessile filter feeders, buried in soft or fixed to hard substrates. A few can swim. **Gastropods** fall into distinct groups: Prosobranchia (mainly the popular shelled species) have their gills in front of the internal organs, Opisthobranchia (most sea slugs) at the posterior end of the body. The highly specialised **Cephalopoda** have the molluscan foot modified into 8 or 10 prehensile tentacles equipped with suckers and/or hooks and arranged in a circle. In its centre lies the mouth with a strong parrot-like beak, used to crack open the shells of other molluscs and crabs, their main prey. They can swim by contracting the mantle and thus squirting a jet of water.

GIANT CLAMS TRIDACNIDAE

Tridacna gigas **Derawan I., Indonesia**

Endangered giant clam
Le: To 120 cm. Di: Entire Indo-Pacific, strongly endangered (gastronomy!), extinct in the Philippines. De: 7-20 m. Ge: Largest tridacnid, smooth shell with 4 strongly projecting vertical folds, siphonal mantle extends over shell valve margin, no tentacles on incurrent siphon, lives as single individuals in seagrass beds and inner reef areas.

Elongated giant clam

Le: To 100 cm. Di: Entire area, widest tridacnid distribution (Red Sea to Polynesia). De: 2-15 m. Ge: Shell variable with weakly projecting vertical folds, one side elongated, with rows of scales which in contrast to *T. squamosa* are near to each other and mostly only present near the upper shell valve margin (lower part smooth). Mantle extends over shell margin, with colourful patterns, tentacles at incurrent siphon, embeds itself partly into coral rubble and sometimes in live stony coral or calcareus substrate, but not as deep as *T. crocea.* Shallow in dense colonies of up to 60 individuals per m².

Tridacna maxima **Singapore**

Scaly giant clam

Le: To 40 cm. Di: Entire area. De: 10-18 m. Ge: Symmetrical, vertical folds strongly projecting, with rows of scales, their distance roughly equals length of an individual scale. Singly in reefs.

Tridacna squamosa **Pangasinan, Philippines**

Boring giant clam

Le: To 15 cm. Di: Philippines to New Caledonia. De: 7-18 m. Ge: Smallest giant clam, shell finely ribbed with flat weakly projecting vertical folds and a large byssal opening, mantle extends over shell margin and has extremely colourful patterns. Shallow, in dense colonies of up to 100 individuals per m².

Tridacna crocea **Sumbawa, Indonesia**

GIANT CLAMS — TRIDACNIDAE

Tridacna derasa **Pangasinan, Philippines**

Smooth giant clam

Length: To 60 cm.
Distribution: Entire area, Indonesia, Philippines, Australia, strongly endangered species (gastronomy), almost extinct in the Philippines.
Depth: 4-20 m.
General: Shell smooth, with 5 or 6 strongly projecting vertical folds, siphonal mantle extends over shell valve margin and has colourful patterns, tentacles present on incurrent siphon, lives in small groups in areas protected from the open ocean by a barrier reef.

Hippopus hippopus **Papua New Guinea**

Horsehoof clam

Le: To 40 cm. Di: Philippines to Vanuatu, extinct in Tonga, Fiji, Samoa. De: 1-6 m. Ge: Shell with numerous slim vertical folds with red blotches, tooth-shaped shell valve margin, mantle brown or olive, singly in seagrass beds, inner reefs. Below: *H. porcellanus,* to 40 cm, rare, endangered (souvenir trade), extinct (Philippines).

FILE CLAMS — LIMIDAE

Lima vulgaris **Banda Sea, Indonesia**

Common file clam

Length: To 5 cm. Distribution: Entire area north to Japan. De: 5-20 m. General: Common in the subtidal on sandy bottoms down to about 20 m. Tentacles at margin of mantle very long, with numerous narrow rings of the same colour. Shell with less than 30 fine radial ribs with short spines.

File clams are not attached to the substrate and able to move freely by rapidly closing their shell and squirting out a jet of water, thus 'jumping' away from predators (crabs).

THORNY OYSTERS SPONDYLIDAE

Orange-mouth thorny oyster

Length: To 15 cm.
Distribution: Entire area, widespread in Indo-Pacific.
Depth: 5-50 m.
General: A common bivalve, the right shell is fixed to hard substrates on sheltered cliff faces, and under ledges, the shells are often covered with sponge and coral growth which obscures the short spines on the shells. The shell snaps closed if carelessly approached and is difficult to photograph open. It has rows of tiny eyes along each shell lip. Centre photo: A dead thorny oyster attached to a jetty pylon is a welcome hiding place for juvenile surgeonfishes. Below: Colour variant.

Spondylus varius **Tioman I., Malaysia**

Spondylus varius **Pulau Seribu, Indonesia**

WINGED OYSTERS PTERIIDAE

Common winged oyster
Le: To 12 cm. Di: Entire area, widespread in Indo-Pacific. De: 2-45 m. Ge: Shells symmetrical, hinge long, straight, 1 small and 1 long process at each end. Juv. with processes considerably longer than width of shell. Shell brown, rough, scaly, often overgrown. Attached with threads of byssus to fire and stone corals, or gorgonarians. Singly or in small groups at exposed places, filtering plankton. Cemented to the shell is a **Cock's comb oyster** *Lopha cristagalli* (True oysters, Ostreidae), both bivalves are covered by a red sponge.

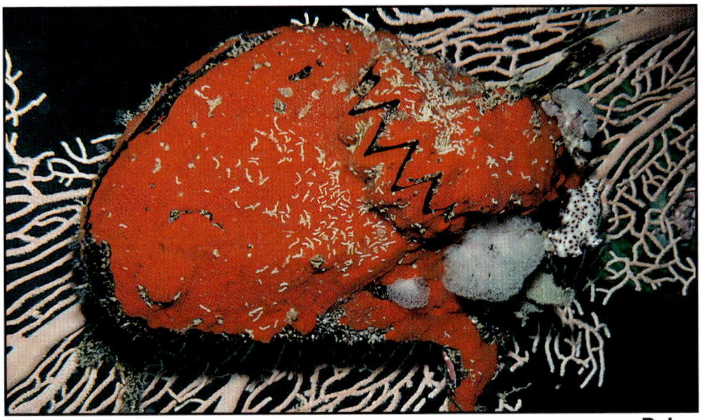

Pteria penguin **Palau**

Strombus vomer Halmahera, Indonesia

Thick-lip conch
Le: To 15 cm. Di: Entire area, north to southern Japan. Ge: This is a rather uncommon and large conch. The shell has a very thick lip bearing brown bars, is cream above and below, irregularly mottled with brown. Below: *S. dentatus,* to 6 cm, entire area, lip with tooth-like process, opening dark violet, uncommon, on sand in reefs.

Lambis chiragra Flores, Indonesia

Finger scorpion conch
Le: To 15 cm. Di: Southern part of area. Ge: The finger-like extensions of the shell are adaptations to a life on soft substrates and also give protection against predators like large fishes. Below: *Lambis scorpius,* to 13 cm, Indonesia, Malaysia, base of first finger with a lobe.

COWRIES CYPRAEIDAE

Mauritia maculifera Luzon, Philippines

Cave cowrie
Le: To 7 cm. Di: Central Philippines, oceanic islands. Ge: Secretive, camouflaged by mantle. In rock crevices, caves, active at night, wandering over the reef to feed. Below: *Leporicypraea mappa,* to 7 cm, Indonesia to Japan. Shell readily recognised by dorsal pattern.

Tiger cowrie
Le: 10 cm. Di: Widespread in tropical Indo-Pacific. Ge: The best-known cowrie due to its size and abundance in shallow coastal reefs. Coloration very variable from heavily spotted and mottled to nearly white. Below: The animal has stretched out its camouflaging mantle over its shiny 'porcelain' shell.

Cypraea tigris Nabucco I., Indonesia

Turtle cowrie
Le: To 6 cm. Di: Entire area. Ge: The living animal of this species looks spectacular: Its jet black mantle is covered with tiny white specks and finger-like papillae. Below: *Lyncina argus,* to 8 cm, entire area, eye-like brown circles on dorsum, varying in size and darkness. 4 black blotches at base near mouth.

Talparia talpa Palau

Isabella cowrie
Le: To 3.5 cm. Di: Entire area to southern Japan, not in Papua New Guinea. Ge: Shell elongate, cylindrical, dorsum greyish with lighter banding and numerous black streaks. Tips orange to black. The specimen in the photo is laying eggs. Below: *Erosaria annulus,* 2.5 cm, entire area, common.

Luria isabella Bali, Indonesia

Ovula ovum **Sulawesi**

Egg cowrie
Le: To 7.5 cm. Di: Entire area. Ge: One of the most common cowries, heavily utilised as decoration in indigenous art in our area, but abundant, often found in very shallow water where they feed on soft corals *Sarcophyton* sp. Distinct from true cowries by teeth on only one side of aperture or none at all.

Primovula roseomaculata **Talaud I., Indonesia**

Red-spotted egg cowrie
Le: To 0.8 cm. Di: Thailand to Indonesia and Micronesia. Ge: Only found on soft corals of the genus *Dendronephthya.* Below: *P. platysia,* to 0.9 cm, Malaysia, dorsal aspect of shell reminds one of the Australian duckbill *(Platypus,* an egg-laying mammal), from which the scientific species name is derived.

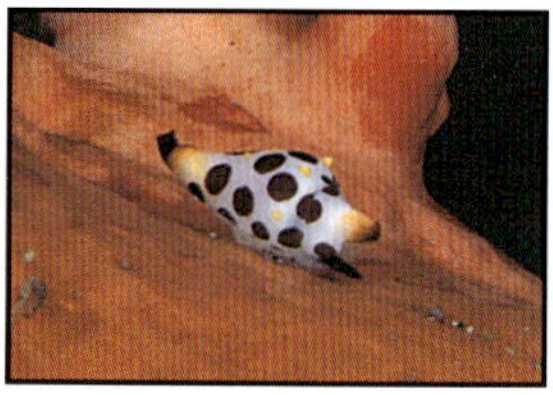

Pseudosimnia marginata **Bali, Indonesia**

Soft coral cowrie
Le: To 1.8 cm. Di: Entire area to Philippines and Japan. Ge: Coloration of red ovate to round blotches on white mantle. Hosts are soft corals of the genus *Dendronephthya.* Below: *Crenavolva tigris,* 1,5 cm, 2-20 m, spectacular shell which is easily discovered and identified. Hosts are gorgonians.

LAMELLARIIDS — LAMELLARIIDAE

Network lamellaria
Le: To 7.5 cm. Di: Indo-West Pacific, southern part of area. Ge: Members of this genus show some sort of coloured network on their dorsum. Below: *Coriocella* sp., to 5 cm, area north to Philippines. Members of this family are often misidentified as opisthobranchs, even in ID books.

Chelyonotus semperi — Sipadan I., Malaysia

HELMET SHELLS — CASSIDAE

Heavy bonnet
Le: To 7 cm. Di: Taiwan to Indonesia. Ge: Shallow, on coral rubble and rocks overgrown by algae and ascidians. Below: *Cassis cornuta,* to 22 cm, largest family member, common on sand.

Casmaria ponderosa — Flores, Indonesia

MOON SHELLS — NATICIDAE

Driller moon shell
Le: To 4.5 cm. Di: In SE part of area. Ge: On clean sand, feeds on other molluscs. Members of this family drill holes into the shells of their prey, inject a digestive juice and suck it out. Below: *Natica lineata,* to 3 cm, Indonesia, Malaysia, eats bivalves.

Polinices aurantius — Singapore

TUN SHELLS — TONNIDAE

Tonna sp. — Cebu, Philippines

Tun shell species
Le: To 10 cm. Di: Known from the Philippines only. Ge: Closely related to *T. perdix* (below), this species is somewhat different in shape and coloration. Below: *T. perdix,* to 11.5 cm, southern part of area. Tun shells are thinner and usually roomier than helmet shells. They are mainly tropical and prefer deeper water.

WENTLETRAPS — EPITONIIDAE

Epitonium billeeanum — Gulf of Thailand

Tube coral wentletrap
Le: To 2 cm. Di: Entire area, tropical IP to the Americas. Ge: Feeds exclusively on corals (genus *Tubastraea).* The yellow and pinkish-red tube coral *T. coccinea* (all tropical oceans, family Dendrophyllidae) grows outside of dense coral populations at current-prone locations. With its long proboscis, the wentletrap feeds on the living coral polyps. Even during egg deposit this female wentletrap continues, the yellow egg mass is deposited in whorls and resembles an extended polyp, a good protection against predators.

MUREX SHELLS — MURICIDAE

Chicoreus palmarosae — Palawan, Philippines

Photo murex
Le: To 10 cm. Di: Entire area to S-Japan. Ge: Uncommon. Striking genus member, coloration varies from orange to dark purple, animals from various localities have brown-stained spines. Below: *Hexaplex stainforthi,* to 6 cm, southern area.

MURIX SHELLS — MURICIDAE

Mud walkman

Le: To 9 cm. Di: Southern part of area. Ge: The distinct body shape of the shells of this genus is due to the muddy areas on which they live. They feed on bivalves. Below: *M. pecten,* to 15 cm, entire area, with three varices per whorl which bear a variable number of long spines. Siphonal canal very long.

Murex tribulus **Bali, Indonesia**

NASSA MUD SNAILS — NASSARIIDAE

Carrion feeder

Le: To 4 cm. Di: Southern part of area. Ge: Moderately common, intertidal down to 10 m. Below: *N. dorsatus,* to 3 cm, common, shell smooth, yellowish. Shallow, on sand.

Nassarius papillosus **Gulf of Thailand**

VOLUTES — VOLUTIDAE

Variable volute

Le: To 8 cm. Di: Malaysia to Philippines. De: 1-20 m. Ge: On mud and seagrass. Shell very variable in colour and pattern. Below: *Melo umbilicatus,* to 30 cm, Indonesia. Animals much bigger than their shells. Diurnally on sand in shallow waters.

Cymbiola vespertilio **Bunaken I., Indonesia**

HARPS — HARPIDAE

Harpa major — **Alor I., Indonesia**

Major harp

Le: To 11 cm. Di: Entire area. Ge: This is the most widespread species of the genus. All members of the family display a relatively small shell in respect to the body size and are therefore not able to retract into it. Below: *H. amouretta,* to 4.5 cm, entire area, found intertidally on sand.

CONE SHELLS — CONIDAE

Conus textile — **Flores, Indonesia**

Textile cone shell

Le: To 11 cm. Di: Entire area. Ge: Widespread, intertidal to 50 m. On coral reefs from the reef crest to deeper waters inside the lagoon, also on sand with or without seaweed, in coral rubble, on mud. Often hidden beneath rocks or coral. Basic coloration white, yellowish-brown flecks are arranged in 2-3 interrupted to solid spiral bands. Preys on many different species of prosobranchs including other genus members. Harpoon-like radula teeth highly venomous, known to have caused several human deaths.

Conus canonicus — **Bali, Indonesia**

Canon cone shell

Le: To 6 cm. Di: Entire area. Ge: Inter- and subtidal coral reef-flats, in sand under coral rocks. Feeds on gastropods. Below: *Conus omaria,* to 8 cm, entire area. Lives in shallow subtidal down to 50 m. Feeds on some prosobranch families including cowries. Venom does not affect mammals, fishes, polychaetes.

AGLAJIDS — AGLAJIDAE

Starry chelidonura
Le: To 5 cm. Di: South China Sea to Hawaii. Ge: Often found crawling on coral during the daylight hours.

The sea slugs of the **subclass Opisthobranchia** are highly evolved gastropods. Most are soft-bodied and many are brilliantly coloured, but all have efficient defences against predators: Some store and reuse the stinging cells from the tentacles of their cnidarian prey; others secrete acids from specialised epidermal cells of the mantle, while many simply concentrate toxic substances from the sponges on which they feed.

Chelidonura amoena — **Nha Trang, Viet Nam**

Pallid chelidonura
Le: To 3 cm. Di: Indonesia. Ge: An easily distinguished species that may have a broken black parapodial band. Below: *C. hirundinina,* to 2 cm, circumtropical, coloration very variable. Can readily be distinguished by the T-shaped marking on its head. Rarely found on sand, preferring rocky substrates.

Chelidonura pallida — **Bali, Indonesia**

PLEUROBRANCHIDS — PLEUROBRANCHIDAE

Forsskals pleurobranch
Le: To 15 cm. Di: Area north to Philippines. Ge: Distinct small, grouped papillae arranged haphazardly, not so in other genus members. Eats sponges, ascidians, sea anemones. Below: *Berthellina citrina,* to 3 cm, entire area, common, smooth.

Pleurobranchus forskali — **Bali, Indonesia**

Nembrotha chamberlaini **Batangas, Philippines**

Chamberlain's neon slug

Le: To 7 cm.
Di: Philippines to Japan.
Ge: Besides feeding on tunicates (ascidians), the species was also observed eating coral polyps.

The **order Nudibranchia** (True sea slugs) belongs to the subclass Opisthobranchia which includes the headshield slugs and sea hares but by far its most-numerous and best-known members are the nudibranchs (literally, 'naked gills': In most species the gills or branchial plumes are outside the body). They occur thoughout all the world's oceans and are present in most habitats.

Nembrotha rutilans **Lembeh Strait, Indonesia**

Root neon slug

Le: To 5 cm. Di: Indonesia to East Australia. Ge: Easily confused with *N. purpureolineata,* which has a yellow band between the purple on the rhinophores, gills and foot and the white body, which is lacking in *N. rutilans*.

Left: Mating occurs between any two mature individuals usually stimulated by seasonal factors. In nudibranchs, the genital organs are always on the animal's right side of the body, a little way behind the head. Like the depicted one, all nudibranchs are hermaphrodites, but it is very rare for any species to self-fertilise.

Nembrotha cristata **(upper), *N. kubaryana*** **(lower)** **Derawan**

Cristate neon slug

Le: To 12 cm. Di: In our area only found in Indonesia, otherwise distributed further west and south in the Pacific Ocean. Ge: This sea slug is always green and black, with a green margin around the base of the foot, green rhinophores, and green pustules on its black body.

Variable neon slug

Le: To 6 cm. Di: Entire area. Ge: Green streaks cover the body. Margin of foot and head, and rhinophores always red. This photo depicts the coloration of the originally described form.

POLYCERIDS POLYCERIDAE

Morose tambja
Le: To 7 cm. Di: Entire area. Ge: Widespread, known for its colour variations. Essentially a subtidal, tropical species occuring near the base of reef slopes. Below: *Thecacera picta,* to 1 cm, entire area. Small, seems to be widespread and shows a variety of different patterns: The lines and spots are very variable.

Tambja morosa **Boracay I., Philippines**

NOTODORIDIDS NOTODORIDIDAE

Minor neon slug
Le: To 9 cm. Di: Entire area. Ge: Yellow and black, feeds on sponges. Below: *N. citrina,* 6 cm, Indonesia. Similar in texture to the previous species, this sea slug is entirely yellow. Its body is more slender, the gills are located at the anterior third.

Notodoris minor **Halmahera, Indonesia**

Serene neon slug
Le: To 10 cm. Di: New Guinea to Philippines. Ge: This species is green with black markings and very large appendages around the gills. Below: *N. gardineri,* to 8 cm, southern part of area. Very variable in depth of brown colour and distribution of patches. Body very rigid and motionless when handled.

Notodoris serenae **Sipadan I., Malaysia**

Gymnodoris aurita **Batangas, Philippines**

Ear gymnodoris
Le: To 9 cm. Di: Philippines to Indonesia. Ge: Known in the Philippines for many years, this species has only recently been recognised as distinct. Orange to red body with whitish tubercles and huge gills distinct. On shallow reefs, hunts other opisthobranchs at night. Below: *G. striata,* to 5.5 cm, Malaysia to Japan. Also eats other sea slugs.

Gymnodoris ceylonica **Irian Jaya**

Ceylon gymnodoris
Le: To 5 cm. Di: Entire area. Ge: This species also preys on opisthobranchs, including family members. Below: *G. rubropapulosa,* to 5 cm, Indonesia. While diving at noon over a flat and sandy ground in only 5 m of water, the photographer found this specimen moving quickly over to a nearby rock.

HEXABRANCHIDS HEXABRANCHIDAE

Hexabranchus sanguineus **Bunaken I., Indonesia**

Spanish dancer
Le: To 40 cm. Di: Entire area. Ge: Most divers are familiar with this mainly red and white patterned snail. The common name refers to its amazing ability to swim with graceful undulations of its mantle margins. Feeds on sponges which make it distasteful.

Coi magnificent slug
Le: To 6 cm. Di: Malaysia to Philippines and Indonesia. Ge: Mantle broad, pale pinkish brown, rhinophores and gills pale brown. Raises and lowers mantle margin while crawling. Below: *C. fidelis,* to 4 cm, southern part of area. Common, with distinct coloration. Feeds on purple sponge *Aplysilla violacea.*

Chromodoris coi **Sipadan I., Malaysia**

Anna's magnificent slug
Le: To 2 cm. Di: Malaysia to Marshall Islands. De: 15-30 m. Ge: With attractive coloration of blue, black, orange, and yellow. Several similar species. Below: *C. willani,* to 2.5 cm, Indonesia to Philippines. Dark blue, gills and rhinophores opaque with white spots.

Chromodoris annae **Sulawesi**

Michael's magnificent slug
Le: To 3 cm. Di: Indonesia, PNG to Philippines. Ge: This chromodoridid is found on reefs covered with featherstars like seen around Alor, Lesser Sunda Is., where the photo was taken. Also found in pairs. Below: *C. hintuanensis,* 1.6 cm, same distribution, S-Japan.

Chromodoris michaeli **Alor I., Indonesia**

Glossodoris rufomarginatus **Komodo I., Indonesia**

Red-margin glossodoris
Le: To 5 cm. Di: Entire area. Ge: Reddish with white and dark margin. Rhinophores and gills are also red and white. Below: *G. stellatus,* to 13 cm, PNG to N-Indonesia. Body firm, black, with white speckles, head is broadest part. Mantle merges smoothly into tail almost as long as body.

Hypselodoris bullockii **Talaud I., Indonesia**

Variable hypselodoris
Le: To 4 cm. Di: Thailand to Indonesia. Ge: Common, range fairly restricted. With several colour variations. It has been proposed recently that this complex of differently coloured forms may actually include several valid species which will require further taxonomic studies.

Hypselodoris infucata **Batangas, Philippines**

Fire hypselodoris
Le: To 4 cm. Di: Our area northward to the Philippines. Ge: Very variable, but always with blue body and red rhinophores and gills. Below: *H. maculosa,* to 3 cm, Philippines to Tahiti. Previously known as *C. decorata.* This species is often encountered hiding among sponges and algal turf.

Malesso halgerda
Le: To 7 cm. Di: Indonesia, Philippines, Marianas. Ge: Recently described from Guam. Photo shows adult and juvenile together, both have the same coloration. Eats sponges. Below: *H. willeyi,* to 7 cm, southern part of area, rather found in mid- to deepwater, has been dredged from 100+ m.

Halgerda malesso **Bali, Indonesia**

WART SLUGS PHYLLIDIIDAE

Varicose wart slug
Le: To 7 cm. Di: Entire area. Ge: Produces a pungent toxin that will kill all life in the immediate vicinity. Has no known predators. Below: *P. ocellata,* to 6 cm, Indonesia to Japan. Very variable, with several colour variants.

Phyllidia varicosa **Flores, Indonesia**

Reticulated wart slug
Le: To 7 cm. Di: Taiwan to Fiji. Ge: Not to be confused with the doridid genus *Halgerda,* this species, like all phyllidiids, does not have a gill plume. Below: *R. fungia*, to 4 cm, Indonesia to Fiji. The absence of pustules and the 'graphically patterned' notum easily distinguishes this genus.

Reticulidia halgerda **Nabucco I., Indonesia**

ARMINIDS — ARMINIDAE

Dermatobranchus sp. — Sumbawa I., Indonesia

Dermatobranch species
Le: To 2 cm. Di: Known only from the Lesser Sunda Islands, Indonesia. Ge: A rather cryptic species which was discovered in coral rubble mixed with some algae. The overwhelming richness of Indonesian reefs holds surprises on almost every dive.

In the family Arminidae (suborder Arminacea) the dorsum (back) has longitudinal striations. External gills may be present or absent. If present, they are found along the sides of the animal, between the notum and the foot. The species feeds on soft corals and sea pens.

FLABELLINIDS — FLABELLINIDAE

Flabellina bilas — Bali, Indonesia

Decorated flabellina
Le: To 2.5 cm. Di: New Guinea westward. Ge: The name *bilas* is a PNG pidgin word meaning 'decoration.' Red bands on cerata and rhinophores distinct.

The Flabellinidae are one of about 20 families in the suborder Aeolidacea. All have elongated, tapering bodies, which bear clusters or rows of elongated processes (cerata) on their back. The cerata contain 'cnidosacs' at their tips to store the undischarged stinging cells ingested from their invertebrate prey (bryozoans, barnacles, hydroids, sea anemones etc.).

FACELINIDS — FACELINIDAE

Phyllodesmium longicirrum — Ambon, Indonesia

Cirrate phyllodesmium
Le: To 14 cm. Di: Indonesia, New Guinea. Ge: Eats soft corals, their zooxanthellae are incorporated into the nudibranch's system to continue producing nutrients by photosynthesis. Below: *P. kabiranum,* to 4.5 cm, entire area.

NAUTILUSES NAUTILIDAE

Palau nautilus
Le: Shell to around 21 cm in diameter. Di: Known only from Palau in the N-Pacific. Ge: This nautilus is very similar to *N. pompilius* (the most abundant genus member in our area) sharing a closed umbilicus covered with a callus. Differs by slightly larger size, fine raised ridges across and along the shell in cross-grid fashion.

Nautiluses live in tropical waters, typically on the steep faces where coral reefs drop into deeper water. During the day they rest at around 300-400 m but at night they can be seen in 'diveable' depths as shown in the photo right.

Nautilus belauensis **Palau**

CUTTLEFISHES SEPIIDAE

Broadclub cuttlefish
Le: To 50 cm. Di: Entire area. Ge: Most common species on coral reefs, at depths to at least 30 m. Capable of many different colour patterns most of which include small white and black spots scattered over the body. Eye typically yellow around edges. Feeding tentacles with around 6 large suckers, many small suckers. Diurnal, capable of excellent camouflage. Eats fish, crustaceans. Cuttlefishes possess a chalky white internal shell known as a cuttlebone. More than 100 benthic family members have been named.

Sepia latimanus **Alor I., Indonesia**

OCTOPUSES OCTOPODIDAE

Day octopus
Le: To 96 cm. Di: Entire area. Ge: Most commonly seen octopus on coral reefs in our area, primarily because it is active during the day. Recognised by the black false-eye spot (which lacks an iridescent ring) and red-brown arm tips with regular rows of white spots.

The family includes the familiar bottom-living muscular octopuses. They are characterised by 8 arms with a double row of suckers without hooks. Brain large, vision acute, communicate via skin. See also **Cephalopods - a world guide** in literature list at end of book.

Octopus cyanea **Palau**

MASTER OF MIMIKRY

Cephalopods make ideal meals. They are composed almost completely of meat, they lack spines or armour, and very few are poisonous. It is therefore in their interests to remain hidden from their many predators. They can do this by using camouflage and mimicry. A master of the latter technique imitating quite a number of diverse models was only recently discovered and has not even been scientificly named yet. Cephalopod specialist Mark Norman comments the photos of Roger Steene.

PHOTOS: ROGER STEENE

The Mimic Octopus (top) and one of its models, a poisonous sole. Photos from Indonesia.

In the molluscan class cephalopods the best camouflage artists are the cuttlefishes, squids and octopuses of shallow coastal waters, capable of dramatic colour and shape changes. They are able to match all sorts of backgrounds from corals to rocks, seaweeds, rubble and sand. Their disguises often combine appropriate colour patterns with raised and branched digits or flaps of skin. These colour patterns often include larger patches of white or black which distract the eye of searching predators from the real outline of the animal.

Some cephalopods take their disguises one step further, beyond simply matching the background. They impersonate whole objects, plants or animals. If caught in the open, many octopuses will impersonate rocks. Some squids impersonate floating pieces of seaweed, while some octopuses move along the bottom with all the arms curled and raised over their body to look like seaweed drifting along the seafloor in the current. *Octopus berrima* has been observed in this disguise timing its movements to coincide with larger swells, as if a stronger wave was pushing drifting seaweed along. Smaller Broadclub Cuttlefish impersonate dead mangrove leaves drifting in the current, complete with stem, ribs and leaf spots such as those caused by insects.

Other cephalopods impersonate animals that are of low food value or hard to catch. Bigfin reef squids *(Sepioteuthis lessoniana)* have been observed to hide in garden eel beds by hanging vertically, stretching and curving the arms above them to take on the shape of swaying garden eels. Their relatives, the Caribbean reef squids *(S. sepioidea),* have been observed sneaking up on prey disguised as parrotfish, complete with "eyes" on the tip of the body and a forked tail (formed by the arms). One undescribed Indonesian octopus impersonates a snail complete with a gliding motion and pat-

Everyone knows lionfishes, but have you ever seen a Mimic Octopus impersonating one of those venomous nocturnal hunters?

terns on the body which look like coloured seaweed growths on the "shell". Mimicry is used in a different way in at least one cuttlefish species, *Sepia apama*, to impersonate the opposite sex.

Recent discovery of a long-armed octopus in Indonesia provid-

This time a banded sea snake (sea krait or sea cobra, *Laticauda semifasciata*) is the venomous model for a Mimic Octopus.

Below: Another amazing mimikry: The Mimic Octopus imitates one of its more distant cousins, a cuttlefish. What's the advantage?

ed the first reports of a cephalopod mimicking poisonous animals. The Mimic Octopus (*Octopus* sp.) lives on shallow sand and mud habitats where it emerges during the day to forage. If approached or threatened it takes on one of a wide range of colour patterns and behaviours that are clearly mimicry of other animals. Impersonated animals include flatfish (soles), lionfish (*Pterois*) and banded sea snakes. Other behaviours appear to be mimicry of crocodile snake eels, stingrays, solitary anemones and even palm leaves. Many of the subjects are either poisonous, dangerous or of low food value. Not only does this octopus match the patterns and form of its subjects, it supplements the disguise with appropriate postures and movements. In the case of the sea snake mimicry, the arms are raised and undulated to simulate snake motion. Similarly, flounder impersonations occur by undulating the wedge of arms as the octopus jets along the seafloor.

The body size of the Mimic Octopus is astonishingly small, only up to 6 cm, with arms to at least 30 cm in length. Its known distribution ranges from the Red Sea to at least New Caledonia. This spectacular long-armed octopus lives in sand and mud flats in shallow waters. It is active during the day foraging for crustaceans and fish that live buried in the sand or mud. It hunts by probing its long arms down burrows or will even worm its way down holes, sometimes swimming through the sand to emerge away from the burrow entrance. It remains buried at night using the holes of other animals as temporary lairs. It often uses the same burrow for days at a time. The common name comes from the ability of this octopus to impersonate the shape and behaviour of other animals, particularly poisonous animals that share the same habitat (see above and photos). But often enough it is up to the spectator's mind which animal might be the model for this master of mimicry which may be related to another unnamed species of Octopus that has the same white "V" on the end of the body.

When peering from the sand, the Mimic Octopus may take on the looks of a sand-dwelling jawfish (family Opistognathidae).

Below:
Although Brittle Stars are found on the menu of several predators in tropical waters, the Mimic Octopus can perfectly play this role too. Again one might ask, why and what for?

CNIDARIANS — CNIDARIA

Coral reefs are probably the best place to see various representatives of the phylum Cnidaria. All are characterised by tentacles armed with myriads of nettling or stinging cells (nematocytes). Each of these cells is equipped with a minute 'safety pin' (cnidocil), which acts as a trigger mechanism. Whenever potential prey like a fish, shrimp, or larger plankton touches the tentacles it cannot avoid bending cnidocils, which in turn leads to an instantaneous avalanche of nematocytes being ejected. This process happens within a few milliseconds, and leaves no chance for escape. Each stinging cell is turned inside out: At first a cover-like flap flips open, then a bundle of knife-like appendages extend, which cut into the prey's skin. Immediately thereafter a long filament from inside the cell penetrates the cut and bores into the victim's tissues while unrolling like a glove's finger. The filament exudes a toxin to paralyse or even kill the prey. Keep in mind that all this is microscopically small and happening a billion times in a few seconds, so there is no chance to get away. One can place a finger on the tentacles of an anemone and feel that they stick to it a little bit. This is caused by the penetration of the filaments into the skin. But make sure before conducting the experiment that it is a harmless species! Two classes of cnidarians are covered here: Hydrozoa (body cavity of polyp without folds) and Anthozoa with the subclasses Octocorallia (nuber eight symmetry of the polyp) and Hexacorallia (number six symmetry). The following pages show representatives of the hydrozoan families Stylasteridae, and the octocorallian orders Stolonifera (Tubiporidae), Alcyonacea (Alcyonariidae to Xeniidae), Gorgonacea (Briareidae to Ellisellidae), Helioporacea, and Pennatulacea.

LACE CORALS — STYLASTERIDAE

***Stylaster* sp.** — **Batangas, Philippines**

Lace coral

Le: To 25 cm.
Di: Entire area.
Ge: Small, delicate coral that fastens to reef walls, usually under overhangs in small colonies. They do not have zooxanthellae and do not require sunlight. The genus contains about 70 different species world wide most are beautifully coloured but it is impossible to determine species identification in the wild and much scientific work remains to be done on taxonomy.

Distichopora violacea — **Bunaken I., Indonesia**

Violet hydrocoral

Le: To 25 cm.
Di: Entire area.
Ge: Grows transversal to strong currents in places protected from sunlight on the reef edge and upper slope. Feeds on microplankton.
Below: *Distichopora nitida.*

ORGAN PIPE CORALS — TUBIPORIDAE

Organ pipe coral

Le: To 100 cm. Di: Entire area, in West Pacific to Vanuatu, Marshall Islands, and GBR. De: 2-14 m. Ge: This unique coral belongs to the order Stolonifera of the subclass Octocorallia. Hence its greyish polyps have eight ('feathered') tentacles. It is found in sun-lit coral reef areas and harbours symbiotic zooxanthellae in its tissue. The calcareous skeleton has a red coloration, the pigment of which is quite stable and a dead organ pipe coral skeleton does not lose its colour. Several spp. described but from dead skeletons only. Regionally over-exploited for jewellery.

Tubipora musica — Flores, Indonesia

LEATHER CORALS — ALCYONIIDAE

Variable finger leather coral

Le: To several metres. Di: Entire area, eastern limit in Pacific is Hawaii. De: 1-40+ m. Ge: This is the most variable soft coral genus concerning shape and size of colonies. They may be low and encrusting, tall and lobed or branching, and flat and plate-like with all possible transient forms. Some colonies grow to only a few cm wide, others cover dozens of square metres when mature. Contracted colonies are hard due to their large sclerites which may be fused into a rock-like substance (spiculite) that may form reef structures. Single colonies may sit on a tall column of spiculite like on a stem. Colonies are brown, yellow, or cream, and polyps are either of the same colour or white or yellow. Living tissue with zooxanthellae.

Family members are the octocorals dominating the reefs of the Indo-West Pacific region. Below: Colonies are resistant to falling dry at low tide (Bali).

Sinularia sp. — Bali, Indonesia

Sinularia sp. — Palau

Sarcophyton sp. **Bali, Indonesia**

Sarcophyton sp. **Cebu, Philippines**

Mushroom leather coral

Le: To 150 cm.
Di: Entire area.
De: 1-20 m.
Ge: The stalked colonies are soft and fleshy. Stalks are short and thick in current-prone locations, but long and thin when growing on soft bottoms or deep in mud. The colonies of some genus members are able to move about at a speed of approximately 50 cm in several months. There are two types of polyps, one up to about 20 mm long when expanded (autozooids), they can be retracted completely; the others (siphonozooids) are small and very numerous, sitting in between the autozooids. The sclerites are mostly well-formed clubs. Colonies are brown, beige, yellow, green, or cream, and autozooid (long) polyps are either of the same colour or white or yellow in some brown colonies. Living tissue with zooxanthellae.
Top left: Several colonies attached to a shipwreck.
Below: Close-up photo of a single 'mushroom'.

Paraminabea sp. **Nabucco I., Indonesia**

Carrot leather coral

Le: To 10 cm. Di: Entire area.
De: Most often below 10 m.
Ge: Uncommon, in dark places (in caves, under ledges) in small groups of a few colonies. These are small and dome- or carrot-shaped. They are only sparsely branched but highly contractile. As in the previous genus there are long, competely retractile autozooids with tiny siphonozooids scattered in between. Sclerites are spindle-shaped or spherical, and usually coloured. Colonies are orange, red, yellow, pinkish-white to cream, and polyps are white or orange. Tissue without zooxanthellae.

Tree soft coral

Le: Mostly to 20 (exceptionally to 200) cm. Di: Entire area, widespread in Indo-Pacific from Africa to Micronesia and Polynesia. De: Most abundant below 20 m. Ge: Regionally common to abundant, found in locations with strong and unidirectional currents, rarely in surge zones. Genus members also in muddy estuaries or deep oceanic. The colonies are highly branched or bushy. There is only one type of polyps with short, stout tentacles. Polyps contain conspicuous bundles (several mm in length) of sclerites which give the colony a spiky appearance and feel. Sclerites are of spindle or irregular shape and strongly coloured. The upper parts of the colonies are brightly red, orange, pink, purple, or white due to the colour of the sclerites; lower parts often pale or white. Living tissue without zooxanthellae. Right: A colony with long, slender branches and retracted polyps. Below: A short-branched colony with fully expanded, bright polyps, photographed at night (Talaud).

***Dendronephthya* sp.** **Bunaken I., Indonesia**

Cauliflower soft coral

Le: To 45 cm. Di: Entire area. De: 15-40 m. Ge: Most soft corals can be identified to species only after collecting and investigating them under a microscope, especially polyp and sclerite structure. Right: A group of rather pale colonies. Below: A single colony.

***Dendronephthya* sp.** **Ambon, Indonesia**

TREE SOFT CORALS — NEPHTHEIDAE

Litophyton sp. — Talaud I., Indonesia

Broccoli soft coral
Le: To 50 cm. Di: Entire area. De: 5-22 m. Ge: Bush- or tree-like, very soft, all polyps of one type, spindly sclerites colourless, colony cream, brown, yellow, purple, with zooxanthellae. Common, in quiet reef zones. In the centre of the photo a colony among various other corals. Below: Another species.

FLOWER SOFT CORALS — XENIIDAE

Heteroxenia sp. — Palau

Leek flower soft coral

Le: To 5 cm. Di: Entire area to Melanesia and Palau in the East, Japan in the North. De: 1-20 m. Ge: Small, slippery (mucus) colonies with one type of often pulsating stalked polyps. Sclerites minute, white. Stalks of colonies and polyps white, cream, yellowish brown, brown. In clear water reef habitats, often common in current-prone areas, in groups of several dozen colonies. Cells contain zooxanthellae (microscopically small unicellular symbiotic algae).

ENCRUSTING SHEET CORALS — BRIAREIDAE

Briareum sp. — Layang I., Malaysia

Encrusting sheet coral
Le: To several metres. Di: Entire area. De: 1-15 m. Ge: Family with one genus only, but species shape and colour very variable: Sheets, knobs, fingers, often magenta. Wide variety of habitats, avoid very clear water. Below: Close-up.

Anthogorgia sp. **Sipadan I., Malaysia**

Blue spiny gorgonian

Le: To 50 cm. Di: Southern part of area. De: 25-45 m. Ge: This sensational photo of an hitherto unknown gorgonian standing between two *Annella* fans, could show an *Anthogorgia.* The growth form and the blue polyps point to this genus, however, the structure of the axis appears enigmatic.

Previous page: A beautiful representative of another genus in this family, *Acanthogorgia isoxya.* Its polyps are spiny. The photo was taken at the coast of Ambon, Indonesia.

GORGONIAN FAN CORALS SUBERGORGIIDAE

Annella mollis **Nabucco I., Indonesia**

Giant fan gorgonian

Le: To 400 cm. Di: Entire area. De: 5-60 m. Ge: Flexible colonies grow in one plane, branched, but not anostomosing (branches connecting netlike with each other). Sclerites long, smooth, partially fused. Stems light brown to dark reddish-brown, polyps white. Without zooxanthellae. In areas with slow currents, also in very turbid areas, often covered with mud etc. Left: With diver behind fan, depth 55 m. Small photo immediately below: Red variant. Bottom small photo: Yellowish variant.

KNOTTED FAN CORALS — MELITHAEIDAE

Knotted fan coral

Le: To 100 cm. Di: Entire area, north to Philippines. De: 20-40 m. Ge: Axis of stem with conspicuously swollen nodes. Highly branched colonies grow in one plane, as large fan or bushy with several parallel fans. Small branches may fuse into net-like maze. Sclerites very variable in shape, spindles, clubs, discs etc. Stems yellow, orange, red, dark purple, pink, white, polyps (one type only) white or purple. Without zooxanthellae. In areas with slow currents, also in very turbid areas, often covered with mud etc. Relatively common under ledges, in crevices, on walls. Right: From 20 m. Below: Indonesia, with crinoids.

Melithaea ochracea — Palau

GORGONIAN BUSH CORALS — GORGONIIDAE

Yellow bush gorgonian
Le: To 20 cm. Di: Indonesia to Philippines. De: 10-25 m. Ge: Bushy, highly branched. Beige to pale brown. With zooxanthellae. Common in currents on clear outer reef slopes. Below: *Rumphella aggregata,* to 100 cm, entire area, 2-20 m.

Pinnigorgia flava — Talaud I., Indonesia

Ellisella ceratophyta **Sipadan I., Malaysia**

Ellisella cercidia **Bali, Indonesia**

Heliopora coerulea **Luzon, Philippines**

Above left: Purple horn sea whip
Above right: Orange straight sea whip
Le: To 100 cm. Di: Entire area, genus very widespread. De: 5-30 m. Ge: Colonies of family members with calcified horny axis, may be unbranched, branched, or in net-like fans. Branching type is important for identification. Genus members highly or sparsely branched, often bushy, with dichotomous branching (splitting into two at regular intervals). Red to orange, pink, white orange-yellow with red polyps, red with white polyps. Without zooxanthellae. Clear to turbid waters, but rare above 5 m. Inshore reefs, in current.

Left: Blue coral
Le: To 150+ cm. Di: Entire area. De: 2-18 m. Ge: The only living shallow water species of the **order Helioporacea.** Living fossil, closely related forms known from the Mesozoic era (100 mya). The only octocoral with a massive calcareous (aragonite) skeleton like hard or fire corals. Skeleton blue (bilirubine), brittle, living colony olive (zooxanthellae). Polyps minute.

SEA PENS — PTEROEIDIDAE

Sea pens
Le: To 60 cm. Di: Entire area. De: 10-300 m. Ge: The Order Pennatulacea (sea pens) is a diverse, but rather poorly known group of octocorals with 16 families at present. The majority of species live in the deep sea, but many others can be found in the shallow waters of our area. They usually inhabit soft bottoms and many completely withdraw into the substrate during the daylight hours. It is because of this behaviour that divers and snorkellers are often not aware of their presence. Sea pens have a large central axial polyp (oozooid) with an internal calcareous axis. Half of it is buried as an anchoring 'root' in the ground and never seen unless dug up. The upper half reaches into the water and catches plankton with numerous other polyps (auto-, siphonozooids) looking like a feather (name). Sea pens often harbour commensal crustaceans among their polyps.

Pteroeides sp. Viet Nam

Pteroeides sp. Batangas

HOST ANEMONES — STICHODACTYLIDAE

Aurora host anemone

Wi: To 30 cm. Di: Entire area. De: 1-5 m. Ge: Singly in cracks on the reef terrace or sandy bottoms. Host to several anemonefish species and *Dascyllus trimaculatus*.

The cnidarian order Actiniaria belongs to the subclass Hexacorallia (with a number six symmetry of the polyp), and comprises many types of anemones found on coral reefs, but relatively few are readily visible. Most are well hidden in cracks or underneath rocks. The most conspicuous ones are those occupied by anemonefishes. These host anemones belong to different families and range in diameter from a few cm to over a metre. The body of all anemones is a hollow tube (column), the closed end of which (pedal disc, foot) usually serves in attaching to a hard substrate, the other one (oral disc) has a mouth opening surrounded by tentacles which contain myriads of stinging cells.

Heteractis aurora Palawan, Philippines

HOST ANEMONES — STICHODACTYLIDAE

Heteractis crispa **Sulawesi**

Leather anemone

Wi: To 30 cm. Di: Entire area. Ge: Hosts various *Amphiprion* spp. Below: *H. magnifica,* to 30 cm, entire area. Colour variable, tentacles greyish with yellowish tips. Exposed on hard substrates of drop-offs and slopes, may move around to change location. Also is host of various *Amphiprion* species.

Stichodactyla haddoni **Alor I., Indonesia**

Haddon's anemone

Wi: To 75 cm. Di: Entire area. De: 4-40 m. Ge: Warty stem in cracks, oral disc carpet-like. Tentacles do not sting, 2 cm long at periphery, 5 cm at disc centre. Below: *S. gigantea,* to 70 cm, entire area. Column smooth, without warts or spots. Carpet-like tentacles strongly stinging. Host to *Amphiprion* species.

TREE ANEMONES — ACTINODENDRIDAE

Actinodendron plumosum **Mabul I., Malaysia**

Feathered anemone

Le: To 40 cm. Di: Entire area. De: 3-20 m. Ge: Strongly resembles certain soft corals, but tentacles long, bushy. Foot anchored in the ground, The animal can retract very fast. Lives singly on sun-lit sand bottoms or coral rubble. At least 15 cm of substrate are needed to dig in and hide.

The most common anemone in our area is the **Bubble anemone** *Entacmaea quadricolor* (family Actiniidae, sea anemones), for this species see the story ANEMONE-CITY on p.168-169.

DISC ANEMONES — DISCOSOMATIDAE

Elephant ear
Wi: To 20 cm. Di: Entire area. De: 1-20 m. Ge: Disc large, usually found in groups of less than ten. Do not produce as many acontia (stinging filaments, not tentacles) as the smaller species, but whether they sting as severly as smaller family members is not known. With small commensal shrimps.

Amplexidiscus fenestrafer — Flores, Indonesia

Disc anemone species
Wi: To 10 cm. Di: Entire area. De: 10-30 m. Ge: Hexacorallian order Corallimorpharia (false corals). Disc-like, tentacles short or absent, column very short. Able to reproduce asexually by budding, thus covering large areas of substrate in a short period of time, overgrowing even living organisms.

Discosoma sp. — Wakatobi, Indonesia

BLACK CORALS — ANTIPATHIDAE

Spiral black coral
Le: To 30 cm. Di: Entire area. Ge: Only the skeleton of black corals is actually black, most are white to yellow, green or brown. Below: *Stichopathes* sp., to 500 cm, entire area, both genera are often mixed up, but sizes are significantly different.

Cirripathes spiralis — Kalimantan

Acropora robusta **Banda Sea, Indonesia**

Robust table coral

Le: To 50 cm. Di: Entire area. Ge: Restricted to shallow reefs where common on reef margins exposed to strong surf.

The stony or hard corals of the hexacorallian order **Scleractinia** are by far the most important reef builders. Millions upon millions of tiny coral polyps live in close symbiosis with even more numerous brownish, unicellular algae (zooxanthellae) embedded in their tissues. It is this symbiosis that enables the polyps of these so-called hermatypic corals to shed calcium carbonate at such a high rate that their skeleton can grow up to 15 cm per year.

Acropora clathrata **Palawan, Philippines**

Solid table coral

Le: To 200+ cm. Di: Entire area. Ge: Anostomosing branches form an almost solid plate.

Individual stony coral polyps resemble fleshy sacks topped by a ring of tentacles around a central mouth opening, very similar to miniature anemones. They sit in calcareous skeletal cases, which are secreted by the polyp itself. Colony members are connected by living tissue, so nutrients available in one section can be shared around. Major groups of corals can be readily identified underwater, but species recognition often is only possible by microscopic examination.

Acropora palifera **Sulawesi**

Pale staghorn coral

Le: To 90 cm. Di: Our area northward to the Philippines. Ge: Another coral of this speciose genus. Below: *A. millepora,* to 40 cm, entire area.

HONEYCOMB CORALS — FAVIIDAE

Bubbly honeycomb coral
Length: To 9 cm.
Distribution: Entire area.
General: Note the extended polyps seen in the large photo at the right versus the retracted polyps seen in the small photo below.

Euphyllia ancora — Kalimantan

CRATER CORALS — DENDROPHYLLIIDAE

Orange cup coral
Le: To 40 cm. Di: Entire area. Ge: Moderately common, in areas exposed to currents, not in caves like other genus members. Below: *T. micrantha,* to 35 cm, entire area, common.

Tubastraea coccinea — Komodo I., Indonesia

MUSHROOM CORALS — FUNGIIDAE

Sun mushroom coral
Le: To 14 cm. Di: Entire area. De: 2-12 m. Ge: Colonies of family members attached or free. Below: **Shield mushroom coral** *Fungia scutaria,* to 17 cm, common, not attached, on reef slopes exposed to surge. Photo: Polyps extended.

Heliofungia actiniformis — Sangihe I., Indonesia

FLAT WORMS — PSEUDOCEROTIDAE

Pseudobiceros bedfordi — **Nabucco I., Indonesia**

Elegant flat worm

Le: To 6 cm. Di: Entire area. Ge: With boomerang-shaped markings, can undulate its margins and swim. A nocturnal species that lives on sand. Below: *P. ferrugineus,* to 5 cm, entire area, relatively common, beautifully coloured.

POLYCHAETE TUBE WORMS — SERPULIDAE

Spirobranchus giganteus **(small),** ***Sabellastarte indica*** — **Mabul**

Christmas tree worm
Le: To 1.5 cm. Di: Entire area. Ge: Phylum Annelida (segmented worms). Has calcareous tube in living coral, two colourful tentacle spirals, operculum.

Indian tube worm
Le: To 10 cm. Di: Entire area. Ge: On sand, family Sabellidae.

Filogranella elatensis — **Bali, Indonesia**

Delicate tube worm
Le: To 50 cm. Di: Entire area. Ge: In colonies, large masses of tubes superficially resemble a coral. Tentacles reveal that they are not polyps but worms.

MERCILESS JAWS

Sometimes, nightmares come to life. At least that's how it seemed when underwater photographer Tony Wu first came across the Bobbit worm. Looking like a miniature replica of a science fiction movie monster this uncanny creature inhabits the muddy bottoms of South East Asia tropical waters but is only rarely seen due its nocturnal habits.

PHOTOS: TONY WU

The Bobbit worm lies in ambush like a living bear trap, poised to snap its deadly jaws shut on unlucky victims. The hair-like sensory organs on its head may act as motion detectors that allow the worm to respond to disturbances in the surrounding water. While I looked on, a hapless bristleworm crawled over a waiting Bobbit like this one - a fatal mistake. The moment of death was lightning quick and impossible to catch on film.

It was during a late night dive at Secret Bay, a shallow, murky bay at the northwest corner of Bali, Indonesia. Adjacent to the terminal for the ferry service between the islands of Java and Bali, the sheltered bay was set in dark volcanic sand. Near shore, the sand was quite fine, gradually becoming more coarse and compact toward the centre of the bay. The sloping bottom reached a maximum depth of approximately twelve metres.

The water was murky and somewhat colder than I expected given the tropical location - about 25°C. People who live nearby ascribed this low temperature phenomenon to the contours of the bay, which keep exchange of water between the coastal area and the interior of the bay to a minimum.

Normally averse to even mildly cold water, I had travelled here to find and study the mysterious Bobbit worm. My unhealthy pre-occupation with this worm started when a friend showed me a picture he had taken. The image was of an alien-like creature reaching out from its home in the mud, like some menacing apparition from a science fiction thriller. It was long and segmented, with skin that glistened and changed colour like oil on water. Its menacing jaws reminded me of a mountain hunter's trap, poised to snap shut on hapless prey.

After seeing my friend's photo, I set out to find out as much as I could on these fascinating animals. I learned that Bobbit worms are multi-segmented polychaetes that were first described by the Russian scientist Pallas in 1788, who at that time assigned them the taxonomic classification *Nereis aphroditois*. Researchers currently identify it as *Eunice aphroditois*. Mischievous divers have given these creatures the colourful common name "Bobbit", a reference to Lorena Bobbit, who violently removed and discarded her husband's penis as revenge for alleged physical and psychological abuse.

Bobbit worms are among the longest worms in the world, reportedly reaching lengths of up to three metres, though specimens of this length are likely to be exceptional. The worms probably reach sexual maturity at much smaller sizes, perhaps at about ten centimetres. Researchers believe that *E.*

Lying prone on the mud and looking through the viewfinder, I was able to get a victim's-eye view of what it looks like just before the worm makes a meal of you.

aphroditois can live for quite some time, maybe as long as fifty to a hundred years.

Like other Eunicid worms, Bobbit worms are benthic, or bottom dwelling. They burrow into coral, mud or rubble substrate, exposing only the top portion of their bodies for feeding. The exposed area has five sensory antenna arranged in a symmetrical radial pattern reminiscent of starfish arms. The worms have two sets of pincer-like jaws, one serrated for cutting. When the sensory antennae detect motion, these jaws clamp shut to capture food. The worms have protrusions along their underside, which provide traction for movement and securing food.

Bobbit worms are omnivorous, feeding on detritus as well as living organisms. The worms are generally nocturnal hunters and live globally in temperate and tropical waters. In the Indo-Pacific, the worms have been reported from waters around Sri Lanka, the Andaman Sea,

This extremely aggressive (probably very hungry?) worm lunged far out of its burrow whenever it sensed the ray of light that was coming out from my torch. Only a few of all the Bobbit worms I encountered seemed to respond to light at all, but this one was by far the most entertaining, giving me a wonderful view of its glistening skin and alien body. With animals like this in our waters, who needs nightmares?

I found a discarded piece of grilled fish in the water one night, and my buddy held it above a Bobbit worm. Not surprisingly, the hungry worm leapt out of its burrow and clamped its vicious jaws on the food. My dive buddy eventually lost the tug of war, as the worm stripped the fish from his hands and devoured it in the peace and quiet of its subterranean lair.

Singapore, Bali, Java and the Philippines. Given the wide distribution and lack of extensive study of these worms, it is possible that several similar species exist under the current *E. aphroditois* classification. Armed with this background information, I set off late at night to meet one of these strange creatures face to face.

Scanning the shallow, muddy bottom of Secret Bay with my dive torch, it didn't take long to find one. I noticed a relatively small set of jaws, about 2.5 cm across, lying flush against the mud. I settled on the mud, hoping to see more of the worm, but the Bobbit worm remained inactive, just waiting patiently in the dark.

My dive torch began to run low on power, so I switched to a fresh one and directed the new, stronger beam at the worm. Much to my amazement and delight, it responded. The worm leapt out of its trance-like state and lunged at my light. I pulled the torch away, and the worm slowly settled back into place. I flashed the beam of light on it again, and it leapt out again, straining for the light and revealing its shimmering segmented body. Despite the seemingly aggressive reaction, the worm kept it jaws open, clearly not mistaking light for food.

I signalled to my dive buddy to come over, and waited as she settled down next to me. The worm had again retreated into its hole, leaving only its jaws and sensory antennae exposed. Just as I was about to demonstrate the interesting behaviour I had encountered, a large bristle worm attracted by my light crawled into view. We both watched in anticipation as it headed straight for the Bobbit's trap.

"Snap!", the Bobbit worm's jaws clamped violently down on its hapless victim, nearly slicing it in two. The worm engulfed the squirming bristle worm, dragging it into the mud. Both creatures of the night disappeared from view for a few seconds. All was momentarily silent, then the mud gurgled and bubbled like a volcano on the brink of eruption. The Bobbit worm re-emerged, spitting up a few leftover bristles. It slowly re-opened its jaws and settled back for the next ambush.

In total, I spent nine nights locating and observing these worms. Some, like the first I encountered, responded to light as well as movement. Others ignored light completely. A few of the worms were vicious, lunging far out of their burrows at the slightest hint of movement. Others appeared utterly disinterested, even in motion directly above them.

There isn't a lot of information available on these fascinating worms, so it's difficult to say with much certainty whether the behaviour I encountered is typical, or whether behaviour varies from place to place, with age, time of year or other considerations. For dedicated muck divers, the polychaete worm aptly dubbed Bobbit presents a golden opportunity to observe and document a little understood marine organism.

LEATHER SEA URCHINS ECHINOTHURIIDAE

Asthenosoma varium **Komodo I., Indonesia**

Coloured sea urchin

Wi: To 18 cm. Di: Entire area. De: 3-30 m. Ge: Coloration white and red to orange, body consists of plates loosely connected by leathery tissue, thus it shape is variable. Found all over the reef, not abundant but common. Nocturnal, omnivorous, feeds on algae, sponges, and sea squirts. Spines often laid together in bundles. The white bubbles at the tips of these spines are very poisonous. Stings are painful and take days to heal, avoid! See also URCHIN HIKERS on p.256-257.

Echinoderms are an exclusively marine phylum with a long fossil record and five major living classes. Feather stars (Crinoidea) are plankton filterers; brittle stars (Ophiuroidea) skim bacteria off the surface, specialised forms filter plankton at night; sea stars (Asteroidea) are mobile predators; sea urchins (Echinoidea) graze algae and animal epibionts from hard substrates; sea cucumbers (Holothurioidea) sieve substrate for microorganisms, organic matter. Below: *Heterocentrotus trigonarius,* family Echinometridae.

TOXIC SEA URCHINS TOXOPNEUSTIDAE

Tripneustes gratilla **Bohol I., Philippines**

Cake urchin

Wi: To 15 cm. Di: Entire area. Ge: A shallow water urchin that has venomous pedicellariae (pincers). Colour variable, likes to disguise its presence by decorating itself with debris. Below: *Mespilia globulus,* to 5 cm , entire area, nocturnal, family Temnopleuridae.

HORNED SEA STARS — OREASTERIDAE

Horned sea star
Wi: To 30 cm. Di: Entire area. Ge: Common, on shallow sand flats and sea grass beds, usually in sheltered habitats, reported to feed on sponges. Horn-shaped nodules sharp or rounded. The sharp spines are not a hazard. Coloration very variable. Below: *Choriaster granulatus,* to 30 cm, common.

Protoreaster nodosus — **Sipadan I., Malaysia**

SNAKE SEA STARS — OPHIDIASTERIDAE

Blue sea star
Wi: To 40 cm. Di: Entire area. Ge: Very common, large, in a variety of habitats, usually shallow but also to 60 m. Colour variable, second most common is olive brown. Below: *L. purpureus,* to 35 cm, entire area, similar to previous sp.

Linckia laevigata — **Batangas, Philippines**

New Caledonian sea star
Le: To 20 cm. Di: Entire area. Ge: Common on shallow reef flats and lagoon bottoms. The white nodules decrease in size towards the tip of the arms where there is an orange tint. Below: **Necklace sea star** *Fromia monilis*, to 12 cm, entire area, colour & pattern distinct, the red disc has pores, in reefs.

Nardoa novacaledoniae — **Sumbawa I., Indonesia**

Comanthina schlegeli **Talaud I., Indonesia**

Variable bushy feather star
Le: To 20 cm. Di: Entire area. Ge: A very common feather star (crinoid) on coral reefs particularly where there are consistent currents. The coloration is highly variable, and the best way to distinguish this sp. from the many other species of feather stars is by the densly packed robust arms, about 130 of them, and size.

Crinoids sit on top of corals and feed on drifting plankton. Careless handling by divers will result in the sticky, but harmless, arms being torn off. Stalked (like many fossil forms) species live in the deep sea.

Comanthina nobilis **Bali, Indonesia**

Noble bushy feather star
Wi: To 35 cm. Di: Entire area. Ge: 80-200 arms, by day curled up, clings to substrate with short lower arms. Below: *Lamprometra* sp. on *Cirripathes* sp.

Comanthina bennetti **Sulawesi**

Bennett's bushy feather star
Le: To 35 cm. Di: Entire area. Ge: Like previous sp. but with 80-100 arms. Below: **Sawtooth feather star** *Oligometra serripinna,* to 10 cm, entire area, with only 10 arms, small and delicate, takes advantage of the feeding habits of soft corals, gorgonians by sitting on them.

Giant basket star

Wi: To 80 cm. Di: Entire area. Ge: This creature hides in coral crevices during the day, emerging at night to perch on top of the reef. There it extends its grasping tentacled arms to feed on plankton. It is very light sensitive and will react to a diver's torch light. Smaller relatives live on soft corals, *Dendronephthya* spp., which always grow in areas of current thus ensuring a supply of plankton. The photo shows a totally contracted specimen, the usual resting pose by day.

Below: *Ophiothrix* sp., to 15 cm, entire area, class brittle stars (Ophiuroidea), species difficult to identify in the wild, associated with soft corals, gorgonians, and crinoids.

Astroboa nuda **Alor I., Indonesia**

Granulate basket star

Wi: To 80 cm. Di: Entire area. Ge: Like the familiar brittle stars also basket stars belong to the class Ophiuroidea but are readily distinguished by their feather-like branched arms. Below: Another *Ophiothrix* species (family Ophiotrichidae, Snake brittle stars). See also small photo above.

Astroboa granulatus **Bohol I., Philippines**

Bohadschia argus **Cebu, Philippines**

Eyed sea cucumber

Le: To 45 cm.
Di: Entire area.
Ge: The eye-spot markings of this sea cucumber are distinctive. If disturbed this, and some other, sea cucumbers discharge defensive white threads, called Cuvierian tubules, from the anus. They are very sticky, can cause intense skin irritation, and should be avoided. Found in shallow lagoons and rubble areas. This sea cucumber swallows sandy sediment at one end then discharges the cleaned - less any digestible matter - sand from the other.

In our area certain sea cucumbers are regularly eaten (left, below) what may cause concern among naturalists because some species already have become regionally scarce.

Fried sea cucumbers **Luzon, Philippines**

Colochirus robustus **Derawan I., Indonesia**

Robust sea cucumber

Le: To 6 cm. Di: Indonesia, Belau, Philippines. Ge: Sometimes in groups on walls in current. Below: **Lambert's worm sea c.** *Synaptula lamberti,* to 10 cm, entire area, Synaptidae. Mistaken for worms. In large infestations on sponges, appear not to harm the sponge. May in fact help the sponge by cleaning off debris.

Sea apple

Le: To 15 cm. Di: Indonesia to Philippines. Ge: The round body of this small, but conspicuous sea cucumber is short, thick, and soft. It has 5 rows of ambulacral feet which are yellow to orange in colour. The oral disc is violet and surrounded by 10 tentacles which are tree-like in appearance and yellow, orange, or red, often with whitish tips. The body is light blue to greyish underneath, changing to pink and red above. Generally, the coloration is very variable.

Habitat are sandy bottoms in reefs, or hard bottoms like walls or drop-offs in a depth of 5-50 m. Often the animals live in groups, sticking together in holes and crevices. While feeding, the tentacles are bent one after the other towards the mouth opening in the centre of the oral disc. There, the planktonic prey, which has been caught on the branches of the tentacles, is removed and eaten. Below: Oral disc with 10 highly branched, tree-like feeding tentacles.

Pseudocolochirus violaceus **Komodo I., Indonesia**

Pear sea cucumber

Le: To 12 cm. Di: Philippines. Ge: Another colourful small species, always banded violet, yellow, and white. Below: A colour variant of *P. violaceus*.

Pseudocolochirus **sp.** **Batangas, Philippines**

STALKED ASCIDIANS — STYELIDAE

Polycarpa aurata — **Ambon, Indonesia**

Golden sea squirt
Le: To 10 cm. Di: Entire area. Ge: A classic tunicate with two openings. If removed from the sea the tunicate will contract and squirt out water hence common name of sea squirt. Solitary or in groups, common, widespread.

Sea squirts (class **Ascidiacea,** phylum chordates) are among the closest relatives of vertebrates, incl. man. All 2,000 solitary or colonial species are marine and fixed to the substrate. Some resemble colourful sponges. An individual resembles a vase with two openings. Inside is a plankton filtering and breathing organ.

URN ASCIDIANS — DIDEMNIDAE

Atriolum robustum — **Palawan, Philippines**

Robust sea squirt

Length: To 3 cm.
Distribution: Entire area.
General: Ascidians have larval stages whose structure relates them to other chordates. There are over 2,000 species which are found singly or in complex colonies attached to the bottom. Many resemble a sponge but can be differentiated since they respond to touch, something sponges never do. This one may be single or in groups of individuals.

Didemnum molle — **Singapore**

Green urn sea squirt
Le: To 3 cm. Di: Entire area. Ge: The previous sp. should not be confused with this similar, very common sp. which is usually larger but with smaller oral siphons, the tiny holes on the sides of the ascidian. Green to brownish colour: Symbiotic algae *(Prochloron* spp.). Below: A different ascidian colony.

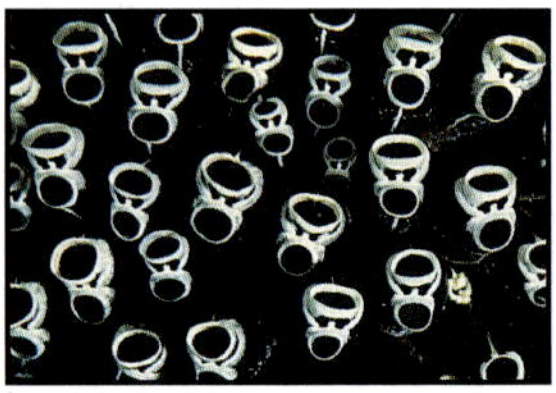

Barrel sponge

Height: To 150 cm.
Distribution: Indonesia and New Guinea to Philippines.
Ge: A huge sponge, the largest in our area, easily identified by its sheer size and very hard, deeply-ridged brownish outer surface. The ridges provide an enlarged surface for the multitude of influx openings. Overall shape variable, often vase- or barrel-like, inner diameter up to 50 cm. Singly or in small groups on coral reef flats and reef walls. Many epizoic organisms settle on the large surface, the inside of the sponge often harbours various crutaceans. Below: An angelfish inspects some crinoids resting near a Barrel sponge's opening.

Xestospongia testudinaria **Alor I., Indonesia**

Elephant ear sponge

Hi: To 150 cm. Di: Indonesia, PNG. Ge: Distinctly fan-shaped, colour variable yellow, brown, green, blue, purple. Sponges are sedentary filter feeders with many small influx and one large efflux opening. There are more than 1,000 spp. in our area, all highly variable in size, colour, and form. Below: *Haliclona* sp.

Ianthella basta **Sangihe I., Indonesia**

INDEX: SCIENTIFIC NAMES

INDEX: COMMON NAMES

BIBLIOGRAPHY
Books and articles used for writing the text:

Abbott, R. T. & Dance, S. P. (1986) Compendium of Seashells. American Malacologists, Melbourne, FL.

Allen, G. R. & Steene, R. (1995) Riff-Führer Indopazifik. Conchbooks, Hackenheim.

Anderson, R. C. (2000) An underwater guide to Indonesia. Singapore.

Briggs, J. (1974) Marine zoogeography. McGraw-Hill Inc.

Debelius, H. (1998) Nudibranchs and Sea snails. 2nd ed. UW-Archiv-IKAN, Frankfurt.

Debelius, H. (1999) Indian Ocean reef guide. UW-Archiv-IKAN, Frankfurt.

Debelius, H. (2000) Crustacea guide of the world. UW-Archiv-IKAN, Frankfurt.

Fabricius, K. & P. Alderslade (2001) Soft corals and sea fans. 264 pp. Australian Institute of Marine Science, Townsville, QLD, Australia.

Grasshoff, M. & G. Bargibant (2001) Coral reef gorgonians of New Caledonia. Paris.

Kuiter, R. H. & H. Debelius (1994) Southeast Asia tropical fish guide. UW-Archiv-IKAN, Frankfurt.

Kuiter, R. H. (1998) Tropical reef-fishes of the western Pacific, Indonesia and adjacent waters. Jakarta.

Lorenz, F., Jr. & Hubert, A. (1993) A guide to worldwide cowries. Verlag Christa Hemmen, Wiesbaden.

Masuda, H. & Y. Kobayashi (1994) Grand atlas of fish life modes. Tokyo.

Myers, R. F. (1999) Micronesian reef fishes. Guam.

Norman, M. (2000) Cephalopods - a world guide. Conchbooks, Hackenheim.

Veron, J. E. N. (1986) Corals of Australia and the Indo-Pacific. Angus & Robertson, Australia.

NOTES:

DO YOU KNOW THE OTHER VOLUMES OF THIS SERIES?

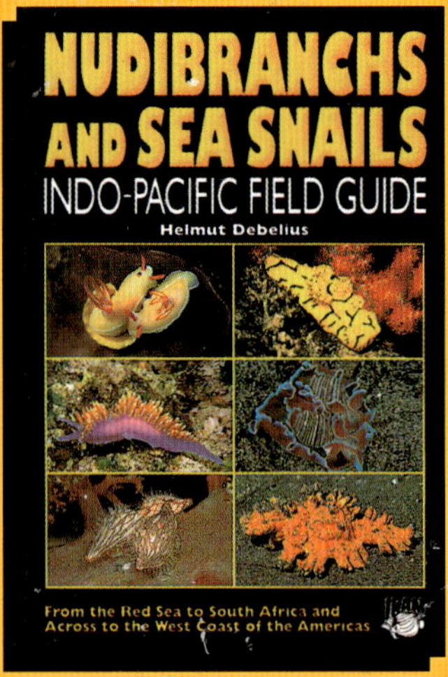

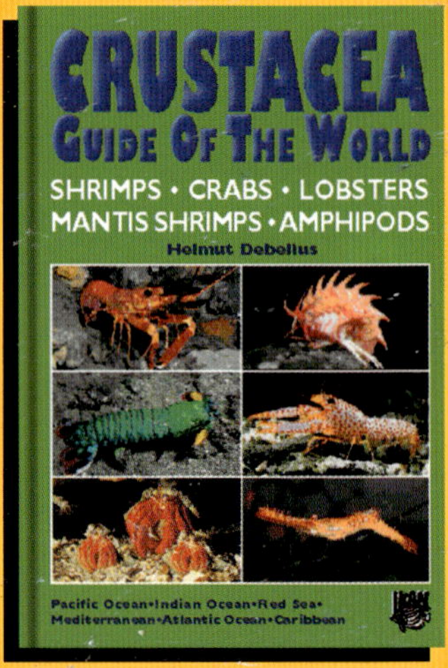

To be ordered directly by CONCHBOOKS
Fax: (Int:49) 671-8963942 or
Email: conchbooks@conchbooks.de